'When it comes to climate ▮▮▮▮▮▮ longer technological, they ar▮ ▮▮▮▮▮ in nature. Read *Turn the Tide* ▮▮▮▮▮▮▮▮▮▮ ▮▮▮▮ and better understand those obstacles – doom, denial and defeatism – and how to circumnavigate them.'

– *Michael Mann, Distinguished Professor of Atmospheric Science at Penn State University and author of* The New Climate War

'Are you losing sleep worrying over the climate crisis? I am. Do you feel powerless when confronted by such a huge global issue? I do. Is climate change starting to affect your mental health? If so, panic not, help is at hand with this brilliant book, which acknowledges that climate anxiety is a real issue and empowers the reader to handle this anxiety and to use to it to make meaningful changes to the world around us.'

– *Professor Mark Maslin, a deeply worried climate scientist and author of* How to Save Our Planet: The Facts (*Penguin 2021*)

'The science is in, the reality is here, and the noble challenge of the climate movement now is to help fellow humanity embrace their eco-anxiety and transform it into action...and build a more vibrant, productive and humane "normal". *Turn the Tide on Climate Anxiety* takes us on the journey to this wild and hopeful alchemy.'

– *Sarah Wilson, journalist and* New York Times *bestselling author of* This One Wild and Precious Life

'*Turn the Tide on Climate Anxiety* is a very interesting and empowering book that has given me new methods for handling and understanding climate anxiety in a comprehensive way.'

– *Nathan Grossman, documentary filmmaker and director of 'I Am Greta'*

'We all have a role to play in addressing climate change. This volume, both highly informed and highly readable, helps to explain the complex and sometimes uncomfortable emotions that accompany that role. Climate anxiety can be an opportunity that prompts us to strengthen our relationship with the earth.'

– Susan Clayton, Professor of Psychology and Environmental Studies at The College of Wooster

'*Turn the Tide on Climate Anxiety* is an important book for an important audience. If you are aware of climate change, aware of what you need to do to address the issue but need a little help with the how, then start here. Galvanize yourself, look after yourself and turn anxiety into action.'

– Joe Duggan, science communicator, founder and coordinator of Is This How You Feel?

Turn the Tide
on Climate Anxiety

of related interest

Emotional Resiliency in the Era of Climate Change
A Clinician's Guide
Leslie Davenport
Foreword by Lise Van Susteren, M.D.
ISBN 978 1 78592 719 5
eISBN 978 1 78450 328 4

The Ultimate Anxiety Toolkit
25 Tools to Worry Less, Relax More, and Boost Your Self-Esteem
Risa Williams
Illustrated by Jennifer Whitney and Amanda Way
ISBN 978 1 78775 770 7
eISBN 978 1 78775 771 4

The CBT Art Workbook for Coping with Anxiety
Jennifer Guest
ISBN 978 1 78775 012 8
eISBN 978 1 78775 013 5

What I Do to Get Through
How to Run, Swim, Cycle, Sew, or Sing Your Way Through Depression
Edited by Olivia Sagan and James Withey
Foreword by Cathy Rentzenbrink
ISBN 978 1 78775 298 6
eISBN 978 1 78775 299 3

How to Get Kids Offline, Outdoors, and Connecting with Nature
200+ Creative Activities to Encourage Self-
Esteem, Mindfulness, and Wellbeing
Bonnie Thomas
ISBN 978 1 84905 968 8
eISBN 978 0 85700 853 4

TURN THE TIDE ON CLIMATE ANXIETY

Sustainable Action for Your
Mental Health and the Planet

Megan Kennedy-Woodard and Dr Patrick Kennedy-Williams

Foreword by Arizona Muse

Jessica Kingsley Publishers
London and Philadelphia

First published in Great Britain in 2022 by Jessica Kingsley Publishers
An Hachette Company

1

Copyright © Megan Kennedy-Woodard and
Dr Patrick Kennedy-Williams 2022
Foreword copyright © Arizona Muse 2022

Figure 4.1 The Climate Emotions Scale has been reproduced
with kind permission from Isaias Hernandez.

The Climate Change Anxiety Scale (CCAS) is reproduced with
kind permission from Susan Clayton and Bryan T. Karazsia.

Figure 5.1 The Five Ds is reproduced with kind
permission from Per Espen Stoknes.

The Signature Strengths Survey is reproduced with kind
permission from the VIA Character Institute.

A CIP catalogue record for this title is available from
the British Library and the Library of Congress

ISBN 978 1 83997 067 2
eISBN 978 1 83997 068 9

Printed and bound in Great Britain by Clays Limited

Jessica Kingsley Publishers' policy is to use papers that are natural,
renewable and recyclable products and made from wood grown
in sustainable forests. The cover is printed on uncoated board to
eliminate the use of plastic in the manufacturing of this book. The
logging and manufacturing processes are expected to conform
to the environmental regulations of the country of origin.

Jessica Kingsley Publishers
Carmelite House,
50 Victoria Embankment,
London, EC4Y 0DZ, UK

www.jkp.com

To scientists and activists.

To the contributors of this book, who shared their stories with us.

To our parents and our children. To our dear friends.

To our beautiful, improbable earth.

Contents

Foreword

By Arizona Muse

Climate change is a relatively new phrase, especially when one considers that until only about 20 years ago it was called global warming. It's something my grandparents did not grow up with, and my parents didn't notice it until they were well into adulthood. One result of this newness is that we are unprepared. Not only unprepared for climate change itself but equally unprepared to handle its effect on our mental health.

In this book, the authors (my dear friends I might add), who are psychologists observing mental health in relationship to the climate crisis, address this pressing issue of our time: how does living on a planet whose climate is changing make us feel?

I am writing this foreword as an example of someone – a concerned citizen, a parent, an activist – who has experienced climate-related mental ill-health and climate-related mental well-health. Several times, I have felt awful because of the things I have learned about climate change, about the plight that our planet is experiencing right now.

I will tell you a story of the first time I cried because of climate change. It was when Australia burned in the fires in its summer of 2019. I had watched videos that day of wild animals roasting, or running away, clearly in a lot of distress, their homeland behind them going up in flames. I watched the accounts of Australian people, often crying themselves, telling the world that the koalas were desperately thirsty, showing families of bats sweating together in the meagre shade of a leaf, until the babies got so hot they died on their mothers' chests of heatstroke and fell to the ground. The distraught scientist who showed this tragic video of the bats pleaded, 'We have to change, we have to do something.' I went to bed that night and cried uncontrollably into my pillow.

That is my first story of grief from climate change. I was sad, I was empty, and I felt powerless because I was so far away. I felt angry as well, angry that the generation before me didn't do anything to stop this. I wanted to blame them; I even wanted them to suffer. They deserved it, after all; how stupid had they been, collectively, to be warned with scientific climate modelling, that if they continued to prioritize the generation of money, the earth would degenerate? I have since felt those feelings rise up regularly, but I have learned that they can help me. They fuelled my desire to push through my innate shyness and become a public speaker on this issue, for example.

These negative feelings also helped me because I wanted to make sure I was right, so I learned. I make sure I learn different things; I choose my education deliberately, and I cross-reference. I must know who commissioned the research, and the most important thing, when learning, is to keep an open mind. I never assume I know more than someone else, because the moment that assumption arises, my own mind shuts off to learning, and I might miss the most incredible surprises. I find the best knowledge I have comes from farmers and indigenous wisdom. Sometimes those two sources overlap, and that's where the golden nuggets rest.

There have been many other times when I feel, and you will feel, overwhelmed. It will seem as if it is simply impossible for us to clean up the earth and change social norms so that they actually serve people rather than the system. Mostly, though, I hang in this delightful zone of action. This is because I know about all the brilliant solutions that are available to us; I see and listen to and read all the brilliant minds who are bravely and courageously moving us into a brighter future, many of whom are women.

Painting it as we go, we all collectively realize that our actions are what will bring it forth. It is up to me how the future develops. It is up to you, too; every thought you have leads you in a direction, so pay attention to your thoughts and your actions and you will see you can use yourself to create the future. This is the deep power in the earth and our human relationship to it.

Ultimately, I will remark on something interesting and unexpected that has come to me from this learning journey, this deep dive into the climate narrative and climate current affairs. In spite of already being a mother, and already having a classifiably 'successful' career, I feel better now than I did before I took up this passion for sustainability and regeneration. I feel happier, more positive, purposeful, driven, dedicated and loving since I started learning about climate change and became an activist. I have overwhelmingly experienced mental well-health.

If someone asked me to choose one single favourite quotation from the book, it's this one: 'Science will save us but not without the stories that engage us.' And the strongest message I can possibly give anyone is this: you are in relationship with the earth. Think about it. On a daily basis, ask yourself: how am I doing with my relationship to the earth?

Arizona Muse is an activist, sustainability consultant and model. She serves as Greenpeace's Oceans Ambassador, Aveda's Global Advocate for Sustainability, and Advisory Board Member for the Sustainable Angle. She is also founder and trustee of Dirt – Foundation for the Regeneration of Earth.

Chapter 1

Turn the Tide on Climate Anxiety

If you've reached for this book, it is likely that at some point, there has been an 'Aha!' or, perhaps more probably, an 'Oh no!' moment.

It's hard to watch the news, scroll through Instagram or listen to the radio without hearing or seeing something disturbing about the climate emergency – especially if it is already on your radar. We might walk down the street and see some litterbug dropping the remains of his single-use packaged lunch on the ground. We might notice someone chatting away on the phone while parked up with the heat on high and the windows down, the car engine pumping away. Over lunch with a relative, they might take the opportunity to inform us that we are under the spell of a conspiracy theory. If global warming is so real, then, 'For heaven's sake how do you explain all that snow in Texas this year?' After all, they saw it on Facebook.

Feeling a bit irked? We witness the impact of climate change on the earth creeping into our lives, sometimes subtly, sometimes catastrophically. A day in January taunts us with its rather delightful picnic weather. Rivers swell into our back gardens. Some of us can't leave our homes because the air is thick with wildfire smoke and a 24-hour news cycle reports that another intense hurricane threatens an imminent arrival. Are we noticing that little (or perhaps massive) undercurrent of anxiety, anger or despair rising?

We, of course, acknowledge and are becoming increasingly aware of the physical effects of global warming, but we're also seeing its impact on our mental health.

At the risk of committing a huge cliché of the psychology profession, can we just ask, when you read that last paragraph, 'How did that make you feel?' Were you thinking, 'Wait, I thought this book was going to rid me of my climate-worry, not antagonize it'? Again, you'll need to forgive us. Mental health is an element of the climate conversation that is often overlooked. As psychologists, we want people to understand that when we are exposed to climate information, we inherently have a psychological response. We are meant to. What makes the work in this book different is that we not only want to ask, 'How did that make you feel?' but essentially, 'What are you going to do about it?' We don't want you to avoid difficult feelings. Instead, we want you to harness them.

We won't pathologize your psychological defences or cognitive biases (don't worry, we'll familiarize you with these terms later in the book), but instead encourage you to turn your attention towards them with curiosity, perhaps even (dare we say it?) 'thank' them, not only for the psychological protection they have afforded you over the years, and indeed will continue to afford you in the years to come, but also for the emotional connection they allow you to have with this beautiful planet. We do this not only because it is good for your psychological health, but also because it can help you to create the necessary power and resilience that will enable you to act against climate change.

Accepting the realities of the climate crisis goes hand in hand with accepting the strong emotions it generates. How we learn to manage these emotions and evoke them to strengthen our wellbeing can also positively impact the planet. This book is not meant to make you feel 'better'. It's meant to make you feel 'stronger' (in a sneaky way, this will likely help you feel better). We want you to feel more able. We want you to feel self-efficacious. We are not here to tell you, 'Oh, don't worry, it'll all work out.' We are here to tell you that you do make a difference, you do have an impact, your emotional wellbeing matters and that the planet needs you to tap into your mental strength. Cultivating awareness is the first step.

Our wonderful human brain – its prerogative is to always be working to protect us, to avoid pain, suffering and unpleasantness. It is inherently pleasure-seeking and risk-averse. We can learn from our pain. It can signal when something feels wrong. We can become appreciative of this, but also kindly challenge it when it is trying to override what serves us. Imagine our favourite dessert: it's right there in front of us – we have a fork in hand. Part of our brain is messaging, 'Eat that now. It's going to be delicious! You deserve it.' Meanwhile, there is another, more complex part of our brain saying, 'Hold, please... You just ate. Remember the diet? You are meant to go for a run this afternoon, and if you are full of German chocolate cake, you probably won't, and then you'll be annoyed with yourself...' Our brains are in constant conflict between short-term gain and long-term reward. Eat the cake now or experience the pride of training for and completing a 10K? One more tequila shot or not regret table dancing in front of your boss in the morning? Avoid the unpleasant feelings that come when thinking about the climate emergency or help the planet today? Override. Don't put it off.

The psychology of engaging with climate change is multifaceted. It requires a high level of self-awareness and openness to stand up to adversity, as well as a willingness to make mistakes, put ourselves out there and, at times, push back against the norm. Ironically, our survival now pits us against many of the psychological mechanisms that, as a species, have kept us safe for a long time. We have this new/old and fast/slow brain telling us how to engage with something really, really big. When we become aware of this conflict and learn to manage it, we recognize our responsibility and the power we have over our own lives. We owe it to ourselves and the planet. It's very much about emotional maturity. It's time to 'adult' and we will actually feel better and more contained for doing so. Mark Manson writes:

This is what's so admirable... The overcoming adversity stuff, the willingness to be different, an outcast, a pariah, all for the sake of one's own values. The willingness to stare failure in the face and shove your middle finger back at it.[1]

When we can stare our emotions back in the face, we can answer them with, 'I feel scared because I care about this. I can deal with feeling scared because I am willing to fight for this.'

There is so much that we can do to help – that is, if we are tackling climate change from this psychologically open place. We can feel honest, empowered and resilient with a commitment to protect what we value. We can open ourselves up and deal with feeling vulnerable. Just as we do when we see children suffering, when we see them raging or regressing or retreating, we don't want to say, 'It's fine. You're being silly. Just stop it!' We want to ask, or better yet to understand, 'Why?' We want to connect with them and validate their emotional experience. As adults, we respond to the same thing. Our needs seldom change that much. We need to feel empathy. Many of us were raised to push against our negative emotions, and although our parents probably had the best intentions, we are going to encourage you to unlearn this. We can retain our emotional maturity while honouring the emotional child within us that just wants to be seen and contained in their struggle to figure it out. Especially since the monster under the bed – in this case, global warming – is actually real.

As more and more people are accepting the science, and as this information garners more attention, the accompanying emotions are obviously becoming salient, and thus increasingly evident in the popular lexicon. As psychologists, we see new terminology and so, too, new ideas. This new language (our eco-emotional taxonomy) includes 'climate anxiety/eco-anxiety', 'eco-rage', 'eco-grief' and more. Their behavioural manifestations have necessarily also generated new definitions. For example, we now understand what it means to 'doomscroll', to 'greenwash'. We know what a 'digital diet' is, and why it is important. These modern incarnations of ancient psychological mechanisms can leave us feeling overwhelmed, detached, hopeless, grief-stricken, angry and despondent, emotions which, if left unmanaged or poorly channelled, may inhibit us from taking necessary, sustainable action to mobilize against the climate crisis.

But wait; we also experience other emotions. We may at times feel excited, invigorated, connected and motivated to support the planet. So how can we turn the tide on climate anxiety? There is good news for you and for the planet: climate anxiety can be alleviated through sustainable climate action – that is to say, action that is sustainable for both ourselves and our environment. Such action, by necessity, involves self-care, self-awareness and balance.

We encourage you to take a moment now to thank yourself for showing up. It is important to acknowledge that you are here, reading this, showing that you are committed enough to be part of the solution. You are making the effort to be present and curious and open. Stay with this for a moment and understand why it deserves your acknowledgment: it means you pushed against many difficult psychological defences in order to have even opened to the first page.

Your impact can reverberate outwards. You can inspire others. Right now, you are 'showing up' for yourself so that you can 'show up' for the planet. Allow yourself those elusive moments of pride for your efforts, protect them and anchor to them when you feel a wobble. You are protecting what you are grateful for, your strength and the planet on which you live.

Incoming waves: the support takes shape

A few years ago, a new client walked through the door of our general psychology practice in Oxford, England, eager for support to manage the stress and anxiety relating to his postgraduate studies. He identified that he was experiencing rumination and debilitating anxiety, which was impacting his mental wellbeing and his ability to work effectively. This person was (and still is) a climate researcher. He was becoming aware that his constant exposure to the harsh realities of climate change was disrupting his life. This was our Client Zero (if he will forgive us the term) – our first client to bring to therapy their struggle in finding that balance between care for the planet and care for the self. He described very eloquently his experience with climate anxiety, and so we include as fully as possible and in his own words, his journey.

VOICES

I would have described myself as a high-functioning sufferer of pervasive climate anxiety. For example, if it was unseasonably hot outside (I can remember a day, for example, in the mid-20s Celsius, or mid-70s in Fahrenheit, in New York in January), I might decide to move my workout to the next day so that I could spend the day inside working and minimize commute time that forced me to go outside and be reminded of how unusually hot it was. Or if it was a weekend, I might postpone plans with friends so that instead of meeting for lunch, we'd meet for drinks after dinner, when the sun had set and the temperatures cooled to closer to normal ranges. When I wasn't able to work my schedule around things that I knew would upset me, I'd still go and do them, but I would struggle throughout with feelings of unease that kept me from being present and engaged.

For me, climate anxiety manifests in one of two ways. There's an initial phase that encompasses and immediately follows the triggering event, and then a second, lower-grade phase that can persist for hours, days or even weeks after the first ends. The best word I can find to describe the initial phase is despair – the feeling of being trapped by something horrific, knowing I have little to no personal agency to resolve the problem and, failing that, that there's not even a viable way to run from it. It feels claustrophobic, like being buried alive. You have no choice but to deal with it.

There are three things that helped me overcome my climate anxiety:

(A) Talking openly about it. Talking about how I was feeling with those close to me was enormously reaffirming, because at no point did I have anyone tell me what I was feeling was stupid or unreasonable. Getting recognition that what I was worried about worried other people, too, that my worries were justified, and that I wasn't alone helped strip away some of the contributing factors (feeling like I had to face climate change alone, worrying about what other negative consequences I might face if I shared how I feel) that made managing that anxiety difficult.

(B) Recognizing that I cannot solve climate change myself, and that I can't stop myself from feeling anxious about it. Climate change is happening. It is a real threat to your wellbeing. It is going to cause you to feel scared; it is rational for you to feel scared, and there's nothing you can do about that. What you can do is stop trying to fight that anxiety when it arrives. Instead of feeling anxious about feeling anxious, you can accept that from time to time you're going to feel badly because you're living through a long, drawn-out, traumatic event that is unprecedented in magnitude across human history. No wonder you feel upset! If you just sit with the anxiety and let it run its course, it becomes a much more shoulder-able load, compared to if you're constantly trying to fight to make yourself feel okay.

(C) Forcing myself to go where the pain is. I started making a point of going outside on the most unusually warm days of the year even if I have no reason to – for example, just to go for a walk. I also started subscribing to regular updates on papers related to the intellectual topics that bothered me the most (e.g. those that study past mass extinction events driven by past climate changes). The more I expose myself to the things that trigger anxiety, the more manageable I find the symptoms become. This is just classic exposure therapy, but it works. Make your anxiety a challenge to run towards instead of something to run from.

Former client (C)

21

This person is no longer a client, but we still work together on a number of projects, mainly around bridging the gap between the science and the emotional dimension of climate research. He has become a great professional success and is making a real impact. He did so by harnessing the power of his emotions to drive his action.

In many respects, his experience was fairly typical of anxiety. However, in other ways, it presented distinct challenges, because it was related specifically to climate change. The climate crisis seems to require a different kind of approach from many other issues that people generally experience when it comes to mental health, not least the question of whether there is anything pathological about the emotional responses that result from global warming. The more we invited conversations about the emotional impact of climate change, the more people would offer their own experiences of the challenges they faced:

- 'I wake up at night thinking about it.'

- 'I feel so guilty.'

- 'It is so frustrating.'

- 'How do I talk to my kids about climate change?'

- 'Our house is in a high-risk fire zone.'

- 'I am doing everything I can and it's not enough.'

- 'There wasn't any snow this year.'

- 'I don't think I should have children.'

They were looking for answers, and although we couldn't tell them, 'Don't worry, it will all be okay,' we also couldn't collude with the idea that the game is up and all is lost. That hopelessness serves no purpose.

Evidence also supported our anecdotal experience. According to a landmark report by the American Psychological Association (APA):[2]

> The psychological responses to climate change, such as conflict avoidance, fatalism, fear, helplessness, and resignation are growing. These responses are keeping us...from properly addressing the core causes of and solutions for our changing climate, and from building and supporting psychological resiliency.[3]

Studies and reports such as this demonstrated that climate change was affecting people psychologically on an increasingly wider scale. This resonated as a huge problem for us because we understand that when we can't process our negative emotions effectively, they can fester and debilitate the actions the planet so desperately needs. We have to keep people working in sustainability, sustained.

And all this before we even begin to acknowledge those who are directly impacted by climate change. The communities facing the reality, rather than the threat, of a warming planet, shifting and increasingly unpredictable weather patterns and climate events. This represents a fundamental issue with (and at times criticism of) the collective conversation about climate anxiety, the disparity in focus on ecological-mental health on the Most Affected People and Areas (MAPA), traditionally referred to as the 'global south', when compared with those areas less directly affected. The 'whiteness' of the problem and of (our perception of) the solution. It also showed us the myopic view we risk holding about the therapies and ideas we offer. A year after writing her own guide to managing climate anxiety,[4] Sarah Jaquette Ray offers a powerful reflection:

> Climate anxiety can operate like white fragility, sucking up all the oxygen in the room and devoting resources toward appeasing the dominant group... Today's progressives espouse climate change as the 'greatest existential threat of our time', a claim that ignores people who have been experiencing existential threats for much longer.[5]

We became conscious to ensure, as best we can, that the focus on climate anxiety did not (and does not) pull attention and resources away from supporting communities already knee-deep in the problem; that privilege does not come at the expense of connectedness and a compassionate, global response; and that the presence of climate anxiety in the climate change conversation doesn't perpetuate an 'otherness' of race or culture, nor does it serve in any way to reinforce white saviourism.[6]

Environmentalism, we increasingly began to appreciate, is truly intersectional. We cannot presume to fully understand the powerful cultural forces at play, nor claim any insight into the 'blindspots' of white privilege. We take special care in the stories we tell and the language we use. For example, when we talk later in the book about the vulnerability of certain communities to the effects of climate change, we do this not to perpetuate an idea that indigenous peoples and communities are inherently 'vulnerable' as a trait *characteristic*, but suggest instead that they share a greater connectedness to the land and water, and that creates a *situation* of vulnerability in terms of livelihood, health and wellbeing. Nor should we assume that the ideas of modern Western individualism (including our psychology) apply universally. For example, there is such power, beauty and connectedness in Native American Ancient Wisdom and other forms of indigenous teaching, much of which can serve as a blueprint for a greater oneness with the natural world. A oneness that perhaps has always been there but has become decoupled by the trajectory of modern societies. However, at the same time, through the conversations we have had in developing this book, we have heard similarities from people around the world (from communities that are more or less directly affected by climate change). Although the psychological language differs, certain tools and ideas felt almost universal: the power of community, of connecting with nature; the importance of activism, but also of self-care and routine.

So, then, we need to remain humble, to listen and to do what we can from the paradoxical position we inhabit in the Western world. We care deeply and yet our carbon footprints are enormous. By our own design, a disproportionate amount of responsibility does fall on the shoulders of those in countries and societies with higher CO_2 lifestyles. This does, however, give us the ability to make a real difference.

Setting sail: our journey into climate psychology

After Patrick's training, he worked for a number of years as a clinical psychologist in a large acute hospital. Much of his work during that time was helping patients and their families to confront challenging realities (chronic or life-limiting illness, or an upcoming surgery). As part of this, they would need to develop real resilience and undergo massive life change, and quickly too, often in the face of situations in which they had little control. The way he came to understand the emotional experience of patients was as being ordinary families facing extraordinary stressors.[7] They would often need to have difficult conversations or make important, life-changing decisions. These were often conversations that families, very understandably, would avoid having, to protect each other. So too the big emotions they entailed. This avoidance would ultimately make the problem feel bigger, and undermine the patient's sense that they could do anything about it, or indeed that they could cope. This overwhelm, in many ways, translates to the psychological experience of climate change. Resilience was also required for doctors and nurses – those on the frontline. The role of the psychologist in a hospital should be, and indeed is, to offer support for these groups too. How to stay well in the face of defeat, having to deliver bad news, experiencing vicariously the traumas of others. Likewise, how to connect with, and allow space for celebrating, those incredible achievements that happen every day in hospitals. Again, this seemed to translate seamlessly to supporting those on the climate frontline, or indeed anyone engaged in the climate emergency.

Megan was working as a coaching psychologist, helping people anchor to their values and set meaningful goals to effect positive change in their lives. She noticed themes and understood that things like procrastination, self-sabotage, inaction, self-doubt, overcompensation, burnout and other barriers would often deter people from achieving their objectives and leading well-balanced lives. Megan saw that motivation and commitment from her clients came from deeply believing that they have the ability to make a difference and the fundamental acceptance that they were 'good enough'. This is what perpetuated their drives to meet targets and achieve more long-term goals as well as overcome setbacks. As Megan began to coach clients about their climate work, she saw that these coaching themes were effective in supporting those with emotions and goals related to climate change. Individuals wanted to be a productive part of the solution. Climate workers wanted to stay active and resilient. Parents wanted to know how to help their children. She recognized early that, drawing from her traditional coaching practice, she couldn't help but kick into the 'notice the feeling, now let's make a good plan' gear, which was beneficial for those suffering from difficult emotions.

We are in this for the same reason as many of you. We are emotionally invested in what we do. We want the people who are working on fixing this climate mess to feel able to carry on (which, we are guessing, in some form or another, will include you). We aren't climate scientists, we aren't sustainability experts, we aren't politicians. But we are people who care. This is a time when we all need to tap into our strengths and lean into what we excel at. We saw that our contribution lies not in the development of carbon-capture solutions or in becoming arrestables for Extinction Rebellion (though a moment away from the kids while schools remained closed under lockdown was nonetheless tempting...). Yet we looked around and saw what was happening and we knew we had to do something. The more people we speak with, the more tell us they feel the same: that they have to do something.

We realized early on that climate conversations needed to be normalized, focused, positive and goal-orientated. We'd notice that people would often become entrenched or fixated on a smaller piece of the puzzle, or perhaps that a setback in their objectives could really knock them off course. They might feel a sense of overwhelm or find themselves distracted, procrastinating or having many projects on the go at once, none of them being as effective as they hoped. This would often feel demoralizing and lead to burnout and doom fatigue. This could lead to the awful spiral of 'I need to do more. What I do is failing. I am not doing enough. I am not enough.' This mentality tends to flatline our goals pretty quickly. While we saw that people could experience this, we also watched as groups would struggle to set achievable, structured, time-limited goals and fail to recognize or reward themselves when they did experience a win. We particularly noted that, within groups, there could be a lot of back and forth, nit-picking, drama and arguing. We saw that these tangents were often the demise of the well-intended outcomes. On the other hand, we also saw how helpful and grounding climate groups could be for people. It was important to champion groups to cultivate tools to stay on topic with clear facts, reveal and allow for emotions, keep values in mind and work together towards measurable, identified goals. These we identified as the logistical antidotes to the paralyzing traps that are easy to fall into.

So we asked ourselves (as we will ask you throughout this book), 'Where are we going to be most useful? How are we best suited to contribute? How can we maximize our climate support?' For us, the answer was that we understand how people think, and, importantly, what gets them moving. So in order to hold space for our own climate emotions, we began to structure a method that would promote mental wellbeing for others on the subject. It felt like our way to help. This felt like our 'Oh no!' moment evolving into our 'Aha!' moment.

We could see where psychological defences (such as avoidance and denial) and cognitive biases were having a negative impact on people, and although, traditionally, we challenge these defences and negative thinking, this didn't feel completely appropriate in a climate context. The 'ailment' (anxiety about climate change), isn't something we can 'challenge'. It isn't an irrational thinking cycle to break, it isn't delusional; it is a scientifically accepted danger to the way of life we know on earth. It's illogical *not* to be alarmed. Likewise, we didn't want our involvement to simply minimize people's emotional experience, to become comfortably numb or complacent in the face of what needs to be done. We didn't simply want business as usual. Yet people need to be contained and supported. We thought that there was space for a shift to occur. There is the opportunity for the crisis to resonate with us, without always carrying the feeling of dread. We don't think people need to unnecessarily suffer.

We got to work. Our mission was to move people from a place of inaction, grief and anger to a place of motivation, power and action.

Our natural tone was one of optimism and connection. We feel that we can engage more people more effectively this way. We need everybody now. It's true that for some people this means doing more, but for many of those who have sought our support, it means doing a little less. Yes, we can be working to do all that we can to stop climate change and feeling awful about it, or we can be doing all that we can to stop climate change and feel good about our efforts. Many we spoke with were already making a massive impact in activism, research or sustainability. It *is* possible to face challenging truths, while at the same time acknowledging, perhaps even enjoying, their journey, their work and the connections they make through climate action, which in turn sustains them.

We began hosting workshops about how to understand and manage the psychological impacts of climate change, and also how to communicate effectively about climate change.

The focus of the workshops was to:

- understand how anxiety affects us and how we respond
- understand barriers to information about climate change
- recognize, name and normalize the emotions people were experiencing about global warming

- identify negative thinking, feeling and behaviour
- develop positive coping strategies
- make plans for action
- pay positive action forwards.

This book will follow a similar structure.

A productive response to global warming requires emotional empathy and long-term, sustained attention with action. This response can be mentally demanding. An additional difficulty is that, often, climate action may not show immediate reward, which is something we humans are rather fond of. We might see the material impact we have after doing a thorough beach clean, but we don't immediately see how denying ourselves a flight to Thailand is making much of a difference. Likewise, our individual actions can feel insignificant when we hear about oil spills and new coal mine proposals. In many ways, global warming is invisible, and unfortunately so too can *seem* the actions we take to battle it. A suggestion posed by psychologist Per Espen Stoknes resonated with us; if CO_2 were a thick brown smog, rather than colourless and odourless, the visibility might have long since generated a more acute global response to climate change.[8] It is even more difficult to compute our impact against this invisible enemy when we are feeling emotionally drained and lacking in psychological bandwidth. We often end up beating ourselves up for what we are not doing. But perhaps more destructively for the longevity of our work, we might beat ourselves up *despite* all the hard work we are putting in. Perhaps we grow apathetic and want to give up.

There is this dichotomy that exists in psychological treatment that goes a bit like 'Yes, accept your negative emotions but also change what's making you feel rubbish and keeping you from what you want.' To some degree, this works well for climate change. Acceptance and change feel dichotomous but, in fact, they can coexist. Indeed, they are two parts of the same process. Although we talk a lot about accepting the negative emotions, we can also acknowledge that feeling positive about the climate work we are doing is allowed too. If our thoughts and feelings are negatively affecting us, we can challenge them and feel better without diminishing our impact on climate change. Positive and deliberate behavioural change eases the emotional stressors of painful climate emotions and works, in hand, to combat global warming.

In the wake of COVID-19: the unavoidable parallels

We wanted to get the message out, but then something happened. Epidemiologists warned we were overdue for a novel disease that would have a global impact. We ignored the threat. Within a week of sitting down to write a book about the biggest crisis the human species has faced, another crisis, COVID-19 began and continues as we type.

Many of our workshops and engagements were postponed in 2020. Understandably, much of the world didn't quite have the collective emotional energy to contemplate two global existential crises at the same time, although the two are intrinsically linked.

Not surprisingly, we saw an increase in contact for psychological support during the COVID-19 pandemic, as people became increasingly affected by quarantine restrictions, existential anxiety, health anxiety, depression, job insecurity and the frontline medical workers' need for a space to process what they were seeing each day. For a while, we paused in our writing to support this need. As it played out, we noticed that the advice we gave for managing climate anxiety bore a striking resemblance to that which we were giving to help people navigate through the COVID-19 experience.

We spoke with climate researchers and psychologists during this time and noted how the early responses to COVID-19 rang familiar: denial, dismissal, pre-trauma, excitement, grief, pontification, polarity and more. If we went down the rabbit hole, it was easy to lose sight of the human connection, to forget about empathy and self-compassion, to struggle to respond in a helpful way, to panic or to bury our heads in the sand. We saw how difficult it was for some to change their behaviour, but also how many did. These are many of the characteristics that we see when people talk or think about global warming.

The general sentiment for the COVID-19 experience seemed to be 'This sucks, we are screwed'. On 31 December 2020, most were feeling a bit like, '2020 – never call me again! We can finally put a pin in it. We did what we had to do, and we are over it.' But that wasn't the end. It will continue until the virus is under control and mass vaccination takes place. One of the difficult things about COVID-19 was (and is) the continual moving of the goal posts; first, second and third waves, lockdowns, hope with the new vaccines, yet no real, tangible end in sight. On a micro level, this is similar to why it is difficult to encourage sustained attention and behavioural shift in relation to climate change. As psychologists, it is our job to challenge that sentiment of 'We are screwed'. This is not a helpful perception.

For example, just look at the mobilization of frontline workers, the dedication and commitment to do everything they could. Look at the innovation and intrepid actions that made real solutions possible. The manufacturing of the vaccines was truly remarkable. It is a great example of how business and government can work together to create game-changing results – and fast. One survey conducted, deliberately, across the sociocultural spectrum in the UK revealed that, over the course of 2020, people in all groups reported a shift in attitudes from 'It's everyone for themselves' to 'We look after each other', with people ultimately subscribing widely to the idea that 'Coronavirus has revealed the best of human nature'.[9] Likewise, with people's worlds shrinking, it brought an increased appreciation of the natural world. In some ways, because people have seen the (admittedly short-lived) environmental benefits of changing human behaviour during lockdowns across the world, but also with restricted movement, people have spent more time immersed in the green or wild areas close to home, bringing a renewed sense of natural connection.[10]

Henry Olsen, in a *Washington Post* feature, offered an insightful opinion on how the COVID-19 immunization initiative (Operation Warp Speed) could translate to the climate crisis from some unexpected participants:

> Operation Warp Speed reduced regulatory barriers and provided a guaranteed market for private companies to do what they do best – innovate and produce. That push and pull gave companies such as Moderna and Pfizer the incentives to get to work. Thanks to this public-private symbiosis, we are now vaccinating hundreds of thousands of people a day less than one year after the pandemic struck the United States. That accomplishment is unheard of, and could not have been done by either the private or public sectors working alone.
>
> A similar approach to fighting climate change could bring Republicans to the table on climate change discussions. 'Operation Green Planet' could combine federal deregulation that impedes the development or use of green technologies, such as nuclear power. It could also use federal purchasing to push private companies to innovate across the board. The policies shouldn't try to pick winners or losers; instead, they should spur competition to discover new ways to reduce greenhouse gas emissions or increase their capture and storage.[11]

Of course, we are mindful and cautious here. We are not encouraging the constant consumer, thriving corporations, individually driven society or business as usual (or back to pre-COVID) mode that has led to the climate disaster; no, but look at how perceptions can shift, and change can happen, with the appropriate motivation – and quickly if the market demands it. Not that we believe capitalism is the answer to the climate crisis. Constant growth has reached its limit in a world of finite resources. But rather the synergy of public and private institutions can, quite simply, work wonders.

The year 2020 brought with it exposure to many broken systems. Now that we have had this experience, first-hand, of a global pandemic, alongside civil rights action, threats to democracies and the evident climate disasters, the reality of life in the time of multiple crises and the toll this takes on our quality of life is starting to sink in, or perhaps bash us over the head. There were Californians on lockdown restrictions because of COVID-19, while at the same time unable to leave their homes because of the poor air quality due to the estimated 4 million acres of land that burned in the same year. Simultaneously, Southern states in the USA braced for massive hurricanes. Globally, similar stories emerged. This has been an incredibly difficult time for many. The compounding weight of the crises we face has a substantial impact on both our individual and collective mental wellbeing. We have started to realize that we are more vulnerable than we had previously thought, but humans are highly motivated to protect ourselves and that which we love. With this often comes a resolution to listen to the facts, commit to actions and make behavioural changes.

COVID-19 also highlighted other broken systems, in terms of its disproportionate impact across gender, race and cultures. We learned that women were more likely to experience COVID-related mental health effects, and intimate partner violence, than men,[12] and that people from some racial and ethnic minorities, due to long-established structural social inequalities, were more likely to be hospitalized from COVID-19-related complications.[13] We saw how richer countries had greater manufacturing and distribution capabilities, and so there was a lack of global equity in terms of access to vaccine supplies.[14] We learned that the reasons for this disproportionality were not in any way related to inherent differences in terms of the resilience across these groups, but because of the wider socioeconomic systems that determined their lives. It was these systems that we began to question – their cracks exposed by a global crisis. It is these systems that we must continue to question, addressing their disproportionality, in our collective global response to climate change.

As we continue to endure the global COVID-19 crisis, we can benefit by utilizing this time as a glimpse into the way the world may respond in a few years' time, if nothing is seriously done about the health emergency that is climate change. We see the importance of tactical communication. Although it has been done for years with global warming, painted at times as a lefty-hippy-woo-woo conspiracy, the degree to which science became political, facts contorted or devalued in the collective mind, in the time of Corona was staggering. But again, this shone a light into the darkness. We learned about the power of facts and of truth. Using this to motivate us, we can also see what needs to be done. Where we saw this lacking in the case of COVID-19, we can learn and adapt communication to steel ourselves to tackle climate change and to support and educate others. We can benefit from seeing the amazing advances made, the human connections and compassion, and remind ourselves that when there is a want, there is a way.

When French President Emmanuel Macron first addressed the nation to inform citizens that France was moving into an extended period of lockdown, he asked the country to 'question the development model in which our world has been engaged for decades'. He said, 'There will come a time in the future where we will have to learn from what we are living through but it is this ability to learn that has seen us come through decades of crisis.'[15]

We have the ability to learn and change course.

How to get the most from this book

There is an increasingly voluminous and important body of literature on 'what' to do about climate change. This book focuses more on 'how'. As psychologists, we know only too well that, for the most part, although people are becoming more and more aware of what they can do and the changes they can make, this insight alone does not guarantee behavioural change. As Charly Cox, founder of the Climate Change Coaches highlights, 'The climate crisis is not an environmental problem. It is a behavioural problem.'[16]

The 'how' in relation to individual and collective climate action is both an art and a science. It requires a keen inward gaze, a little psychological understanding and plenty of perseverance. Patrick remembers one particular seminar he attended that brought this idea wonderfully to life. He and his cohort of trainee psychologists were asked to stand up, then remain standing if they had managed to heed certain well-known physical health recommendations over the past month. 'Stay standing if you've consumed fewer than 14 units of alcohol each week.' A few courageous souls sat down. (Or maybe they passed out. Who can tell?) 'Stay standing if you have eaten five portions of fruit and vegetables each day.' A few more sat. 'Thirty minutes of moderate exercise per day?' 'Seven hours or more of sleep per night?' More still, with a little exasperated laughter this time – we were in the middle of writing a doctoral thesis, after all. By the end of the exercise, there were precisely no people left standing.

The point was that we can know what is good for us – these were all guidelines we were familiar with – but this itself didn't correlate well with our health-related behaviour. This knowledge/action discrepancy among us 15 trainee psychologists (a population, it has to be said, that is generally on the attentive end of things) was representative of the greatest challenge facing public health, and it is a behavioural not technological one. We sometimes call this the 'last mile problem'. How do we elicit behavioural change when the solutions are staring us in the face? The scientific community continues to provide an unprecedented understanding of human-caused changes to our climate, and environmental charities, NGOs and public figures likewise offer, in plain terms, what we need to do. In exactly the same way, our 'last mile' climate problem is behavioural. The rest has been done already.

When writing this book, we have drawn upon a number of theories and applied disciplines, from evolutionary psychology to behavioural economics, cognitive psychology, psychodynamic theory and coaching. We delve somewhat into the field of climate change communication because it's helpful to gain insight into what is happening in our minds, in order to normalize and accept the psychology that comes with processing global warming. Understanding our psychological defences will promote self-awareness. We will then explore the personal values that anchor us and the narratives about our role in climate action. We will work on cultivating abundance and resilience, and we will make wellbeing plans, as we set goals to enable behavioural change. Finally, we will look beyond with aspirations of our own ripple effects turning into waves of change.

We have included activities and exercises in the book, and we encourage you to interact with these. Evidence tells us that these programmes are much more effective if you take the time to pause and practise each exercise. Try not to read through the book all in one sitting, but rather take the time to set realistic and meaningful goals and embed the ideas and exercises into your daily life. Practise the meditations or find some that resonate for you. Try to commit to making them a part of your daily routine. Each chapter builds on those preceding it, so we advise you to read them in order.

Allow yourself some frustrations. Behavioural change – even seemingly modest changes – can be challenging at first. We can often revert to old habits of thinking and doing without even realizing. So notice this and forgive yourself along the journey. Most importantly, celebrate your successes at each stage. The magic, they say, is in the doing.

Finally, we have been blessed to be able to include personal accounts, provided to us specifically in the writing of this book, to help illustrate people's own experiences of the intersection of climate change, psychological resilience and action. These accounts, some of which are anonymized and others not (at the choice of the individual), are given by our former clients, collaborators and other figures in the sustainability movement, interchangeably. However, what unites their contributions isn't necessarily the work they are doing (although they are mostly actively involved in the earth sciences, ecology, activism or sustainability in some way), but rather their focus on the psychological journey they are on. These contributions take the form of the *Voices* sections throughout.

We are in equal measure both humbled and truly thankful for these contributions to our book, and also for the work these people continue to do in the fight against climate change. Our hope is that their stories will resonate with you and offer a sense not only that you are not alone in your emotional experience but also that there are people out there who have found meaningful ways of turning the tide. Their voices are real, human and powerful. They are humble and have, in distinct yet universal ways, acknowledged their strength and fallibility. They open up to their struggles and have shared what they have found to work best for themselves. They speak to the emotional experience of loving their planet, and they are representative of a much wider, ever-expanding and incredible movement of people, around the world, fighting for what they love. In the sage words of young climate activist Xiye Bastida:

> It's time to change our mindset toward implementing solutions. A vibrant, fair, and regenerative future is possible – not when thousands of people do climate justice activism perfectly but when millions of people do the best they can.[17]

Chapter 2

What We Know About Climate Anxiety (and What We Don't)

Grist writer Miyo McGinn declared 2019's biggest pop-culture trend to be climate anxiety:

> 2019 was officially the year the climate crisis went mainstream. Think about it. It's not a niche worry for small pockets of concerned citizens. Instead, the climate crisis began appearing in movies, songs, and books meant for widespread consumption – a bar that many of them actually met.[1]

As we were starting out, the terms 'eco-anxiety' and 'climate anxiety' were just beginning to enter the wider lexicon. They have now become media buzzwords and relatable forms of describing how we feel about the state of the planet. As McGinn noted, climate change influences everything from pop music charts to Burger King menu options. More recently, some researchers are beginning to consider the differences between eco-anxiety and climate anxiety. For example, as Panu describes, eco-anxiety can be seen as 'any anxiety which is related to the ecological crisis', whereas climate anxiety can instead be described as 'such anxiety which is significantly related to anthropogenic climate change'.[2] There is also a suggestion that perhaps *fear* (as relating to a specific, identifiable threat) is a more accurate descriptive term than *anxiety* (which instead relates to the unknown and uncertain) in how we conceptualize our worries about the planet. As the definitions are not yet definitively ascribed, we use the terms eco-anxiety and climate anxiety interchangeably in this book.

However, regardless of definition, climate anxiety and eco-anxiety have given people a framework to understand and connect their experiences with others. This has laid a foundation and opened up the dialogue at international and cross-cultural levels.

How widespread is climate anxiety?

Over the past few years, we have relied on a number of questionnaire studies, conducted around the world, to better understand people's level of concern regarding climate change. The results of these continue to support our anecdotal experience of how climate change is affecting people's psychological wellbeing, most notably younger people. For example:

- Among 10,000 young people from 22 countries, 41% said climate change was one of the most important issues facing the world, making it the most commonly cited issue globally. (Ipsos Mori Survey on behalf of Amnesty International, 2019)[3]

- Among American teenagers, 57% said that climate change made them feel scared and 52% said it made them feel angry, both higher rates than among adults. Only 29% of teens interviewed said they felt optimistic. (*Washington Post*–Kaiser Family Foundation poll, 2019)[4]

- A recent British survey suggested that young people were the most affected across the age ranges: 34% of the British public reported feeling anxious because of the environmental emergency; 29% described feeling overwhelmed; this figure rose to 40% amongst younger people aged 16–24. (Triodos Bank survey, 2019)[5]

- In an Australian study of young people aged 7–25, roughly 96% consider climate change to be a serious problem, with 89% saying that they are worried about the effects of climate change. Feelings of disempowerment were significant, with more than 70% of the participants concerned that people do not, or will not, take their opinions on climate change seriously. (Millennium Kids survey, 2019)[6]

Although much of the research has focused on Western European, North American and Australian attitudes towards climate change, one recent large-scale study demonstrated the extent to which people from 50 countries around the world identified climate change as being a 'global emergency'. Although Western Europe and North America had the greatest percentage of respondents agreeing with this statement (72%), others weren't far behind (65% of respondents in Eastern Europe and Central Asia, 64% in Arab States, 63% in Latin America and the Caribbean, 63% in Asia and the Pacific, and 61% in sub-Saharan Africa).[7]

However, questionnaire data carries with it certain limitations, and it has only been in the past 12 months that we have seen validated measures of climate anxiety emerging in the research literature. Using the Climate Change Anxiety Scale (CCAS) that you, as the reader of this book, have the opportunity to complete in Chapter 3, data shows that between 17% and 27% of study participants experienced climate anxiety sufficient to affect their daily functioning in some way.[8] In one recent survey of adults in the United States, roughly one-third of respondents reported 'a little climate anxiety' and around one-quarter reported 'a lot'.[9] However, we still have no large-scale national or international data using tools such as the CCAS to accurately measure the prevalence of climate anxiety. Likewise, we are not able to make any comparisons between age groups, across cultures and so forth. We hope that this will become clearer in the very near future because it represents an important area, particularly cross-culturally, as we continue to grapple with the question of how globally representative our understanding of climate anxiety indeed is.

Is climate anxiety a mental health condition?

At the beginning of our work on this subject, this question seemed to dominate the conversation, and it was gaining a lot of media attention. We found ourselves engaged in a debate surrounding whether climate anxiety is real and, if so, whether it should be classified as a mental health condition.

Certainly, speaking with people who have experienced climate anxiety, many have said that they do not wish for it to be considered anything pathological from a mental health perspective, but instead they believe it is a normal response to an existential threat. This sentiment is echoed in recent work by Pihkala Panu, who describes: 'It seems that many – probably most – forms of eco-anxiety are non-pathological, which causes the need to be careful in both health care and public discussion about the standard definitions and "treatments" of eco-anxiety.'[10] This is an important point because how we decide to classify climate anxiety (as a mental health condition or not) would have implications for how services are established, and if/how future psychological treatments would be funded and delivered.

Often referred to as the 'bible' of psychiatric decision-making, the *Diagnostic and Statistical Manual of Mental Disorders*, 5th Edition (DSM-5), produced by a task force representing the American Psychiatric Association, is the most commonly used system for classifying, diagnosing and treating mental health conditions. Published in 2013, the DSM-5[11] replaced its predecessor, the DSM-IV-TR ('text revision'), which was published 13 years previously. This means that although the DSM-5 is considered a 'living document', in reality, it is only updated periodically. Climate anxiety (or related terms such as eco-grief and eco-anxiety) is not recognized by the current DSM-5 classification. Formally, therefore, this means that no, climate anxiety is not officially considered to be a mental health condition, although this may change in future editions of the DSM. In order for it to do so, however, much more research will need to be done to understand how, and if, climate anxiety is sufficiently distinct from other forms of anxiety, to warrant its own diagnostic label – and, likewise, if there is a justification for doing so.

We are often asked whether climate anxiety is really just anxiety with a different focus (climate change). This raises an interesting question about the classification of mental health. There are absolutely overlaps. However, if, for example, a person were to seek help due to specific fears and unhelpful patterns of behaviour relating to their health, they would likely be told they are experiencing health anxiety/hypochondriasis (or, as the DSM-5 describes, 'illness anxiety disorder'). Likewise, if instead they describe unhelpful thoughts, feelings and behaviours in relation to social events, it would likely be considered to be social anxiety (or 'social anxiety disorder' in the DSM-5). Because of decades of research, we have been able to understand the differences in how these problems can manifest themselves, and what keeps them going. This has led to tailored and specific psychological therapies. For example, cognitive behavioural therapy (CBT) programmes for health anxiety would look different to CBT programmes for social anxiety, and so forth. It is therefore important to consider differential diagnoses in these cases, to make sure the right approach is being used.

Our point here is not that we necessarily believe that climate anxiety should be formally classified as a mental health condition at all, but instead that, by continuing to investigate how it affects people, we will likely become better able to adapt our therapeutic approach. The more we know, the better we can support.

Whom does climate anxiety affect?

The 'direct' and 'indirect' psychological effects of climate change

One will often see the literature on climate change and mental health split into two phenomena. The first involves the direct mental health impacts of climate events. Commonly, these include studies conducted in the aftermath of floods, wildfires and other natural disasters linked to climate change. In cases such as these, there are known localized increases in depression, anxiety, posttraumatic stress disorder (PTSD) and substance abuse following these events, and even, the literature suggests, in domestic violence.[12, 13] For these direct events, we know that children and young people are particularly vulnerable to the psychological effects. Likewise, because of the social disadvantages of women in parts of the world, climate migration (following flood, drought and so forth) can make them more vulnerable to sexual exploitation or trafficking, or more likely to work in the sex trade to support their families whose livelihoods have been lost due to climate events.[14]

Direct climate events often bring with them indirect, 'secondary' consequences. Alongside forced climate migration, unemployment and loss of home and community are common.[15, 16] These, by eroding people's sense of place and connection, lead to further psychological consequences. For example, rural communities in Ghana were forced to relocate to the capital city, Accra, because their farming practices were no longer viable due to changes in their local conditions, attributed to climate change. This migration was reported to have left these communities feeling nostalgic for their homes, sad and hopeless.[17] Similar 'ecomigrations' have occurred in communities in India, Nepal, Bangladesh and elsewhere around the world, and are predicted to increase in frequency. Other temporary migrations have occurred – for example, the displaced people in the aftermath of Hurricane Katrina in the United States or wildfires in Australia. When we are forced to move from our home, this can lead us to question our identity and purpose, which are both important for resilience.

The second indirect category relates to the psychological impact of hearing about, discovering or otherwise engaging in the wider climate crisis. Climate anxiety, therefore, would fall into this category, although it is important to bear in mind that these two (direct or indirect effects) are not always completely separate from each other. For example, in the Pacific island nation of Tuvalu, regarded as being particularly susceptible to the effects of climate change, researchers found that people were experiencing psychological distress not only regarding local effects of climate change but also in a wider sense, from hearing about climate change on a global level and its potential implications. The latter is more reminiscent of how we might define climate anxiety, therefore illustrating that in communities exposed to the direct impacts of climate change, 'climate anxiety' may coexist at a greater degree than in other parts of the world.[18] There is also the strong argument that as unprecedented weather events become more commonplace around the world, the distinction between being directly or indirectly affected by climate change will become less clear, and examples such as the Tuvalu case study become more widespread.

There is, however, another argument to be made. Increasingly, studies are framing the research question differently, looking instead at resilience (rather than psychopathology) in communities following climate events. Some report that, actually, rates of resilience are higher following direct climate events than, say, anxiety.[19] Likewise, researchers are curious to understand the best predictors of community resilience in these scenarios including, for example, high levels of social capital and strong community leadership, regardless of the socioeconomic makeup of the local population.[20]

Young people

As shown earlier in this chapter, many studies point to younger generations as being particularly vulnerable to the effects of climate anxiety. In the UK, a recent BBC Newsround survey found that 17% of children and young people experience some form of sleep or eating disturbance due to worries about climate change.[21] One suggested reason for this is the fact that young people are at a key point in their physical and psychological development, and thus are inherently more vulnerable to the effects of additional stress.[22] Also, of course, it is the younger generations who have more to lose in terms of future climate events. They are deeply invested in the future of the climate crisis. Perhaps for this reason, young people are more likely to rate the impact of climate change on their lives as higher. In one survey, 51% of people aged 17–34 think global warming will pose a serious threat in their lifetime, compared with 29% of those aged 55 and over.[23] In perhaps the largest and most international study to date, 10,000 young people aged 16–25, from ten countries across both the global north and global south, highlighted the extent of distress regarding the climate crisis. Forty-five per cent of those taking part described that climate distress and anxiety were impacting on their daily lives and functioning. Seventy-five per cent reported believing that 'the future is frightening' (which rose to a staggering 92% in young people in the Philippines). Levels of distress were typically higher in poorer countries, or those more directly affected by climate change. Additionally, this study directly identified that a crucial driver for this distress was government inaction.[24]

This fits into a wider picture in which the rates of psychological distress are higher in younger people, and, with these rates increasing alarmingly since the early 2000s, in young adults aged 18–25.[25]

There are high levels of engagement in younger people, too. As many as 52% of young people describe themselves as 'motivated', and 24% of them had already taken some form of action against the climate crisis (either having attended a rally, taken part in a school walk-out or written to a local politician).[26] There is, quite simply, a huge potential, and appetite, in young people to make a substantial difference.

Earth scientists/sustainability workers

Other groups that are more vulnerable to experiencing climate anxiety fall into a broad category comprising earth sciences professionals, researchers, scientists, sustainability advocates and activists. Not surprisingly, climate anxiety is more common in people who care about environmental issues.[27] This is one explanation for why those working in the climate space may be more vulnerable, as they very much fit into this category. Likewise, the toll of communicating a constant stream of bad news, as well as needing to engage with those sceptical of the effects of climate change, might explain why these groups are more at risk. This unmet need was documented in rather eloquent fashion by Joe Duggan's project *Is this how you feel?*[28] which, since 2014, has been compiling handwritten letters sent by climate scientists documenting their emotional responses to climate change. Although not a 'randomized' or representative sample, of course, the project highlights the range of strong emotions experienced by this group. In the United States, initiatives such as the Adaptive Mind Project are underway to develop an infrastructure of psychological support for earth sciences professionals.

This group is interesting, though, as what the limited research available seems to indicate is that they may be more vulnerable to climate anxiety for these reasons, but also that they have, in many ways, a greater level of psychological resilience, in terms of coping with being exposed daily to the harsh realities of climate change. In the case of activists, one suggested source of resilience stems from their proficiency in using social media to form connections around the world.

VOICES

Activism isn't always happy.

I am so grateful for my friends; having friends who know. You don't have to explain it. They know. And it's funny in this online world – we will just call each other and be quiet together.

It's like a form of solidarity... And these are friends that are across the globe. My best friends are from India and the Netherlands and we've never met. We only see the top half of each other's bodies but there is this connection because we understand this pain and grief.

Mitzi Jonelle Tan, climate justice activist from the Philippines

Other predisposing factors?

When we have looked at the variables that might predict a person's psychological response to climate change, there are also several other potential factors to consider. Many studies described a worsening of pre-existing mental health difficulties following climate events, meaning that there is some suggestion that people are at greater risk of climate-related distress if they have a history of other psychological difficulties. This has also been seen in people who were not necessarily exposed directly to localized climate events. For example, people with a pre-existing diagnosis of obsessive-compulsive disorder (OCD) have reported increased climate-related checking behaviour.[29]

We also need to consider the predisposing factors as not necessarily being about the individual, but also, arguably to a greater extent, the wider context and systems in which the person exists – the sociocultural factors that determine a person's or community's reaction to climate change, including psychological responses such as climate anxiety. As climate psychologist Professor Susan Clayton explains:

> As with other social issues that affect mental health, such as sexism, racism, and poverty, we must find a way to respond to individual problems without losing sight of the social consequences – to talk about climate anxiety as a psychological experience without implying that the causes, and appropriate responses, are intrapsychic.[30]

This becomes particularly pertinent when we consider the intersection of racial justice, women's rights and global inequality, and how these relate to our individual and collective experience of the climate crisis.

This is notably true for young people, who report that they feel ignored, abandoned and betrayed by adults and that governments are failing to respond appropriately.

Though we encourage you to support yourselves through your climate emotions towards personal goals for climate action, we want you to know that we fully recognize the flaws that exist in the wider systems. Impetus from government must drive change. We can use our individual actions to put pressure on these systems to change. Acknowledging the response (or lack thereof) from corporations and governments can help us to harness our senses of injustice and anger and demand through divesting, voting and applying political pressure to make our voices heard. It's okay to feel let down, infuriated and disillusioned by these systems. Feel these things, talk about these emotions, but we return to the question; 'what are you going to do about it?'

Receiving climate messages

This is not a book about climate science. This is a book about psychological resilience and adaptation in response to climate science.

We want everyone to talk about climate change. We want people to spread the word about how climate change makes them feel and what they are doing about it. Although we want to remain well informed, we also want people to understand that they do not need to be climate experts to talk about climate change. This can hinder conversations as many people feel they aren't knowledgeable enough or sufficiently informed and thus avoid the topic. Equally, others may feel they know too much, experiencing discomfort in their role as an expert. We also may form an aversion to talking about it, as we find it triggering, or out of concern for others.

Here we look at some climate facts and we want to be mindful of how we are receiving them. We want you to notice if familiar emotions begin to emerge. These facts might be old news to you. You might be reading this book because you are extremely well versed, perhaps a climate scientist who helped bring this information into public awareness. You may be feeling consumed by all the bad news. Or it may be that you are just becoming aware of the problem, and you may not have seen these facts before. In any event, it is helpful for us to connect to some of the kinds of information we may be exposed to as we go about our day-to-day life.

Then we will begin to look at how we may be interpreting these facts, potential barriers to responding helpfully to these facts, coping mechanisms we may employ, and also how to evaluate the results of our actions.

Therefore, let's take a look at some basic climate facts so that we are clear on what we understand about climate change, and how it makes us feel.

Facing the climate reality

- The global population increased from 2.5 billion in 1950 to 7.8 billion in 2020. This huge increase in the number of humans also correlates with an increase in the amount of energy we use each day in our modern lives. This has led to the amount of carbon dioxide in the atmosphere increasing during that time at a rate 400 times faster than at any other point over the past 5000 years. Levels of greenhouse gases are higher now than at any time in the past 800,000 years.[31]

- We have presided over an increase in atmospheric carbon dioxide of over a third since the industrial revolution. This is due to human activity.[32] Consequently, the earth has warmed 1.8°F (1°C) since 1850.[33] This has led to warmer temperatures across the globe, retreating sea ice, melting glaciers and a rise in global sea levels, but also increased humidity, which can lead to problems like exhaustion and heatstroke as our bodies are less able to cool by sweating.

- This is also already causing extreme weather events such as heatwaves, wildfires, drought and floods, many of which have not been seen on such a scale since recording began.

- As a result of human-caused climate changes and habitat destruction, species are struggling to adapt. Today's extinction rate is estimated to be hundreds, if not thousands, of times the baseline rate.[34]

- We have fundamentally changed life on earth. Wild animals now constitute only 3% of the total weight of land mammals on the planet. Livestock constitutes 67% and humans 30%. Ten thousand years ago, wild animals accounted for 99.95%.[35]

So these are some of the facts. Just as we asked you to take a quick 'emotional pulse' at the beginning of this book, now is another opportune time to notice how you are feeling. Notice if any emotions may have arisen for you when you read those facts. Again, we are trying to connect to the feelings that this information evokes. We invite you to take a few moments to write down what it felt like to have read these climate facts. What thoughts did you have? What feelings? What physical sensations are you experiencing in your body? Remember, there are no 'correct' emotional responses to this information. You may find yourself feeling alarmed, or even anxious. Equally, though, you may notice a numbed or subdued response. You may feel motivated. Whatever it is, take a moment to engage with those thoughts, feelings and physical sensations below.

My thoughts:

1. . . . .

2. . . . .

3. . . . .

My feelings:

1. . . . .

2. . . . .

3. . . . .

Sensations in my body:

1. . . . .

2. . . . .

3. . . . .

Okay. Deep breath. Maybe do a walk around the room, make yourself a tea or coffee, then return (but please do return!).

Now for some further facts:

- In 2020, American people in seven states voted for climate-positive policies that ensured a total of $2.2 billion for conservation projects in their areas.[36]

- In the decade 2009–2019, renewable energy grew at an average annual rate of 13.9% and continues to grow exponentially in the world's major carbon-producing countries such as China and the USA. Renewables were the only energy type whose growth was in double digits during this period.[37, 38]

- Sales of electric cars have surged, relative to combustion engine vehicles, fuelled in part by tough government resolutions to ban the sale of petrol and diesel cars (in Norway, by 2025, and the UK by 2030, for example).[39] In 2020, one in six cars sold was a pure electric or hybrid vehicle, signalling a huge increase in their popularity and market share.[40]

- The Great Green Wall, an African-led initiative to plant a green corridor across the entire 8000km width of the continent, is already 15% underway. It will be the largest living structure on the planet and is considered a new wonder of the world.[41] As well as rewilding an area three times the size of the Great Barrier Reef, the project is also creating food, water and health security, and increasing gender equality across some of Africa's poorest nations. In January 2021, the Great Green Wall received an additional $14 billion of international funding.[42]

- A similar project in Pakistan, the 10 Billion Tree Tsunami Program, has provided over 60,000 jobs to local people (primarily day laborers) who were laid off as a result of the COVID-19 pandemic.[43]

- Humpback whale populations off the coast of Brazil have increased to an extent not seen since before whaling began in the area.[44]

- As of March 2021, over 14 million people have taken part in Fridays for Future/School Strike for Climate protests in more than 7500 cities, across all the continents on the planet.[45]

- Humans, as with all earth's life forms, are constructed out of elements that were created in stars across the universe.[46]

- You are made of stardust.[47]

Now let's sit for a moment and take in the positive climate initiatives that are happening all around us. Let's think about the actions and innovations, both personal and large-scale shifts that we are seeing each day. What does this feel like?

My thoughts:

1. ....

2. ....

3. ....

My feelings:

1. ....

2. ....

3. ....

Sensations in my body:

1. . . . .

2. . . . .

3. . . . .

Bear in mind that the facts have remained the same. Nothing has changed. However, your subjective experience may have done, as you processed the first set of information, then the second. Global warming is still a clear and present danger, but our thoughts and views about the possibilities that exist to fight climate change may have shifted, even slightly, when we acknowledge all that is being done about it. When our thoughts reflect the possibility of positive contributions, our emotions match these and give us that sense of resilience we need. Our anxiety levels recede, and we are better equipped to take climate action.

The world is not black and white. We can't live a full life without accepting the negative. It is not all rainbows and David Attenboroughs. We must accept that there are indeed some dev-astating realities about climate change, but we can also accept that solutions exist. When we think in terms of 'nobody is doing enough', we forget about all the millions of people who care and are taking action. There are positive initiatives taking place, and we can choose to be a part of the remedy, rather than being consumed by the problem.

When we start to look with curiosity at how our cognitive processes engage when we hear and think about climate change, we can find the strength to be inquisitive about our reactions and harness them to manage our wellbeing. We notice that emotions are vibrations in our body, and like waves rolling in and out from the shore, they come and go. We can look at them with self-compassion and as a way of revealing our deeper commitment to the health of our minds and the planet.

Chapter 3

Navigating Our Responses to Climate Change

Understanding normal responses to threat

As humans, we have a tendency to pathologize anxiety and other strong negative emotions. Or at least, their uncomfortable nature means we try to push them away, to get rid of them as soon as possible. This is understandable. Anxiety doesn't feel good. However, it's important to appreciate that, from an evolutionary perspective, there is nothing pathological about anxiety; indeed, it constitutes part of an essential in-built survival mechanism we share with all mammals and other classes of animal. It's been essential for our survival, but sometimes – as in the case of climate anxiety, for example – it serves us to recognize when it arises and how we can manage it in order to not feel overwhelmed or consumed.

Here we take you on a whistle-stop tour through the world of anxiety, borrowing from discoveries in evolutionary psychology, cognitive neuroscience and neurophysiology. Understanding anxiety won't necessarily make it go away, but it's an important step in putting yourself in the driving seat. You can then develop the skills to respond, rather than react, to anxiety.

Understanding the fight/flight/freeze (FFF) response

Anxiety can be a frightening experience. It's highly physiological. Its effects can be felt throughout the body, but we often don't attribute physical sensations to 'mere anxiety'. It's actually quite amazing how anxiety can manifest itself physiologically, from tingling and numbness to headaches, butterflies and a racing heart. There's a lot happening when we are anxious. The important thing to remember is that, in the short term, this process is perfectly natural and causes minimal harm in the longer term.

The first thing that occurs when we are confronted by a stimulus (be it a news story, message from a friend or a car backfiring) is the amygdala decides whether a threat has been detected. The amygdala is our brain's mini fire alarm. It is a subcortical structure of the limbic system, deep inside our temporal lobe. If the amygdala detects a threat, it signals to the hypothalamus (our brain's relay centre) to trigger the secretion of our stress hormones. Because this process is so primitive, it means that it doesn't adjust itself too well to distinguish between real, physical threats and hypothetical ones. This means that we find ourselves experiencing a similar neurochemical response to worries and times of stress as we do times we are in some form of physical danger.

Because the amygdala also plays a role in memory consolidation (the process of an event moving from our short- to long-term memory), our memories and emotions can become intertwined, and the more 'emotionally activating' an event is, the more likely we are to remember it, and it can trigger a similar stress response when we recall the events. This is an important process in conditions such as PTSD. Put simply, we remember what scares us.

Once the hypothalamus has communicated the need to activate the stress response, the autonomic nervous system (ANS) and hypothalamic–pituitary–adrenal (HPA) axis become activated, and epinephrine (adrenaline) and cortisol are released into the bloodstream. When this happens, we can experience a range of effects in the body. From head to toe, it can look like this:

Dizziness/light-headedness

If we don't use up the extra oxygen in our bodies (e.g. by running or fighting), we can be left with an excess. This isn't harmful but can result in dizziness or light-headedness. People often misconstrue this as an indication they are about to faint. However, this is highly unlikely, as the FFF response actually raises blood pressure slightly, whereas fainting occurs following a drop in blood pressure.

Racing thoughts

The FFF response actually leads to temporary changes in our thinking, as well as our bodies. Thoughts become more rapid, to enable us to make quick decisions in a crisis. However, in a non-crisis situation, or a more long-term or existential crisis such as global warming, this can lead to thoughts racing in a less controllable manner and prove to be less effective in helping us manage our physical response.

Visual changes

Changes in vision are designed to make you better able to spot danger, by becoming more focused and acute. In everyday situations, however, this can result in blurred or 'tunnel' vision. This can feel unusual and disconcerting.

Dry mouth

As the entire digestive system slows or shuts down temporarily during activation of the FFF response, the effects can be felt throughout the body. Even the mouth, part of our digestive system, can become dry, as the whole system is deprioritized to allow the body to focus its energy on the key muscle groups.

Breathing changes

As the body attempts to gather more and more oxygen into the bloodstream to power the muscles, our breathing changes. This often results in shallow, rapid breathing. This in turn can make us feel breathless, which can often be misconstrued as a sign of serious physical danger. When we think we can't breathe, this can make us more anxious, acting as a vicious cycle that inhibits us from moderating our response.

Racing heart

Again, as our body attempts to push more oxygen-rich blood to the key muscle groups, our heart rate increases. This is not dangerous in any way and serves us very well in sports, or situations in which we do need to respond in a hurry. However, this can also leave us with an uncomfortable palpitation feeling in other, non-direct threat scenarios. If we are paying particular attention to a change in our heartbeat, again we can feel as if we are under immediate physical threat.

Muscle tension

In order to help gear you up to either fight or flee, muscles around the body become tense. You may notice, particularly if you are sitting down or staying still, that these 'activated' muscles may shake or tremble. This is a symptom of the FFF response that can be particularly noticeable for people, and annoying if they are, say, trying to take an exam or parallel park in their driving test.

Sweaty palms

We sweat in order to cool the body. This is important to prevent overheating, especially if we are needing to respond to a physical threat. Needless to say, we still experience sweaty palms when the FFF response responds to imagined or psychological threats.

'Butterflies', cramps and stomachache

As noted above, the result of a slowing down of the digestive system following activation of the FFF response can be felt around the body. A common experience is sensations in the stomach. These are very often reported by children and young people when feeling anxious, but they are felt in adults, too. They can take the form of 'butterflies', stomachache or cramping.

Bladder urgency

Rather inconveniently, the FFF response can also occasionally trigger the bladder to empty (or, at least, trigger an urgent need to urinate) as the muscles become more relaxed.[1]

My FFF response

Everyone will have different experiences of the FFF response and at different times. We may feel one, several or all of these responses when our system is triggered. Learning what our individual responses are will help us to gain insight into what we need to do when this system is firing, and the steps we can take to identify an imminent threat or existential threat that requires us to calm this system before we can take meaningful action to mitigate it.

Take some time to identify your own FFF response. Do any of the above symptoms sound familiar to you? If so, in what situations, or following which triggers?

Take the test: Am I experiencing climate anxiety?

Climate Change Anxiety Scale (CCAS)[2]

The following questionnaire was developed by Susan Clayton and Bryan T. Karazsia. On its publication in 2020, it marked an important step in our ability to measure climate anxiety. As you read through the statements below, allocate a score for each between 1 and 5.

Remember there are no right or wrong answers, and usually the first answer that enters your mind is the most accurate.

Please rate how often the following statements are true of you.					
Never 1	Rarely 2	Sometimes 3	Often 4	Almost always 5	My score
1. Thinking about climate change makes it difficult for me to concentrate.					
2. Thinking about climate change makes it difficult for me to sleep.					
3. I have nightmares about climate change.					
4. I find myself crying because of climate change.					
5. I think, 'Why can't I handle climate change better?'					
6. I go away by myself and think about why I feel this way about climate change.					
7. I write down my thoughts about climate change and analyze them.					
8. I think, 'Why do I react to climate change this way?'					
9. My concerns about climate change make it hard for me to have fun with my family or friends.					
10. I have problems balancing my concerns about sustainability with the needs of my family.					
11. My concerns about climate change interfere with my ability to get work or school assignments done.					
12. My concerns about climate change undermine my ability to work to my potential.					
13. My friends say I think about climate change too much.					

Once you have scored your answers, you can also separate them into two categories. Your scores on questions 1–8 represent the 'cognitive-emotional impairment' subscale. This relates to the degree to which your thoughts and emotional experience are affected by climate anxiety. Questions 9–13 represent the 'functional impairment' subscale. This instead relates to the impact that climate anxiety is having on your ability to conduct your daily activities as you usually would.

My cognitive-emotional score (questions 1–8) = ...

My functional impairment score (questions 9–13) = ...

Notes about my scores (a space for you to make any observations about your scores):

. . . .

. . . .

. . . .

. . . .

What can I do with this information?

Each individual statement in the CCAS tells you something about your own psychological responses to climate change. The two impairment subscales were created because the researchers discovered that people seemed to experience climate anxiety most notably in terms of their thinking and emotions, or how it affected their behaviour (or both). This information can be used to help you better understand your individual response to climate change. Does it seem to resonate more in terms of your thinking? Or perhaps you find yourself preoccupied with the question of *why* you are experiencing these thoughts or feelings to this extent. Or, instead, what you notice more is the way climate anxiety affects your ability to work, have fun or connect with people. Maybe you experience variations in these on a day-to-day basis.

Remember that these scores should not be interpreted as meaning that you necessarily have an 'anxiety disorder', but instead can be interpreted to show how climate change is affecting your mental wellbeing at the present moment. If you are concerned about any of your scores, you can always visit your doctor to discuss your psychological health.

You might also use your responses to help create goals for yourself. Were there any answers that you would like to have been different? If so, how? You can then return to the questionnaire after you have finished this book, having given yourself some time to put the ideas into practice. Has anything changed? Do your scores look different?

Chapter 4

Climate Emotions

As we mentioned in the opening chapter, one thing is clear: there is a range of emotional responses that we all experience in relation to climate change, and these need not be pathologized in any way. There has been an interesting philosophical debate around the role of emotions in climate change, centring at times around the Fridays for Future movement. Two arguments[1] formed around the functions of emotions:

1. On the one hand, are emotions an obstacle? Do they detract from our ability to engage in rational thought?

2. Or instead might an emotional response to the climate crisis be the only one that is rational? Would it instead be irrational to be emotionless in the face of global warming?

What is more, how do emotions intersect with climate behaviours? The short answer is that they play a crucial role, but we are still only beginning to understand how. For example, how can we harness the power of negative emotions like worry, without it leading to fear or paralysis? Or, conversely, how can we use hope and connection without causing complacency? One thing is for sure: strong emotions can and do lead to climate action, in the right circumstances. Even guilt has been linked with an increased likelihood of engaging in pro-environmental behaviours such as petition signing.[2] Compassion, too, increases support for good climate policy.

The task, of course, is to develop a greater understanding of your emotional response, when it serves you and when it takes over things in a less helpful way. When it motivates or demotivates, inspires sustainable action or perpetuates inaction. This is what the following chapters aim to support you with. Isaias Hernandez, environmental educator and creator of QueerBrownVegan, created a concise scale of climate emotions, which we believe demonstrates as good an example as currently exists of the range of emotional responses we may have to our planet and its changes (Figure 4.1). It serves as an excellent foundation to better understand and befriend our climate emotions.

Figure 4.1 Climate Emotion Scale: Talking About Earth Emotions
© *QueerBrownVegan*

Hernandez draws together disparate terminology from the past few decades, in part from the groundbreaking work of Glenn Albrecht, pioneer of a lexicon of words to describe our emotional responses to our planet, both positive and negative. In his 2019 book, *Earth Emotions: New Words for a New World*, Albrecht shines a light not only on the wide-ranging and powerful negative emotional responses – *solastalgia* (or a melancholy we experience at the destruction of the planet) being perhaps the most commonly used – but also of those positive emotions, too. This latter vocabulary of optimism, Albrecht suggests, is a powerful tool, which could even help to bring about a new human era that may usher us away from the Anthropocene (the current epoch in earth's history, in which human activity is considered the most powerful influence over the climate and natural world). Instead, Albrecht invites us to contemplate the next epoch, which he calls the 'Symbiocene', where we can become truly connected and intertwined (in our thoughts, feelings and behaviour) with the planet.[3] Climate emotions such as *Eutierra* (a feeling of oneness with the earth) and *Soliphilia* (love for the earth) represent this connection (or re-connection) with and love for our earth, and thus form a wonderful and important set of emotions to pay attention to and harness.

Climate anxiety

A little climate anxiety can be a good thing.

This is something we have heard a number of people say, and with good reason. Many people feel that a certain degree of anxiety about climate change, at an individual and collective level, is a necessary driver for action.

In relation to his own climate anxiety, a client once described to Patrick the need for 'the right amount of disruption'. He explained this as being a sufficient amount of anxiety about climate change so as to continue to motivate him to do what he can, but not so much that it became paralyzing or affected his psychological wellbeing. In a wonderful metaphor, this client likened this disruption effect to the survival of coral reef ecosystems, which actually thrive in relatively unstable underwater conditions involving submarine discharge, which produce increases of certain nutrients such as phosphorus, accelerating their growth. However, too much water disruption becomes harmful. This balance is a careful one. This client saw his relationship with climate anxiety in much the same way. The right amount of disruption (or anxiety) kept him motivated, energized and goal-orientated. At times, however, anxiety would take over and become unhelpful.

Other clients have described something similar – that, essentially, 'the right amount of anxiety can be a good thing'. This means that there are potentially ways that we can harness anxiety, or at least our response to anxiety, to allow us to positively engage in meaningful action. In our workshops for parents and educators, we use a similar metaphor in an aim to evoke the same idea: of 'enough grit to grow', whereby a plant propagated in overly soft soil, without any obstructions, can develop a weaker stalk and ultimately collapse. Likewise, though, too much grit in the soil environment is likely to be damaging to the plant's development. In this goldilocks of metaphors, the solution is 'just right'. Enough grit to develop a resilient stalk, able to grow past the obstacles that exist, and prosper.

VOICES

Back as a student, I read a memorable paper published in *Science* in the 1970s exploring just why ecosystems like coral reefs and rainforests are so diverse.[4] It proposed that it's because the ecosystems are always getting disrupted enough that they never reach an equilibrium, but that all the species that live in them are able to adapt, flourish and persist to the rates of wider change just enough, and with enough space, to keep evolving.

If they got disrupted too much, of course, they can collapse; but get disrupted too little, and they'd lose some of the endless forms of beautiful biodiversity they are home to. So, in other words, the ecosystems as a whole evolved because they got the right amount of disruption.

For me, I wondered how far this metaphor goes. How much does it apply to the climate system, to our societies, our economies, our communities and to ourselves? How do we keep evolving, and keep responding to the 'right amount of disruption'? How do we avoid getting too comfortable? How do we prevent collapse?

As far as I'm aware (and last time I looked at the literature[5]), the trick is to have enough cooperation. There will always be disruptions to our environment – hopefully, not all quite as heavy as the disruptions to the carbon cycle, climate system, biodiversity and other planetary boundaries over the past century – but there will be other disruptions, and risks. But, hopefully, we can cooperate with each other (and by each other, I mean the other inhabitants of Earth) enough to respond to these disruptions, to slow them down, halt them, and 'ultimately' restore whatever dynamic balances are possible in the years, decades and centuries ahead.

I do wonder if the same thing might apply to mental health. The more we support each other, and the more we are able to respond to the disruptions to our minds, to evolve positively from them, to give space where it's needed, and to avoid our own 'irreversible tipping points', perhaps the more we can support a more safe and just future for all, and to helping life on this planet to thrive and flourish, and ultimately not be extinguished before its time.

Former client (D)

The question, therefore, is what does that look like for you? What is the right amount of disruption, or grit, in your life, to keep you motivated, to allow you to grow? What are the signs that you are harnessing anxiety in a productive way and staying resilient? Likewise, what are the warning signs that the disruption is becoming too great? Are there certain thoughts, emotions or behaviours that are suggestive of this for you? Being mindfully aware of these is an important way of maintaining the right amount of disruption. Helping to navigate these questions is more specifically addressed in subsequent sections, but it is a good time to start thinking about what this balance looks like for you.

Climate grief

VOICES

I imagine there are real parallels between climate anxiety and the emotional experience of losing someone close to you – an initial emotional shock when you hear the news, an immediate period of despair when you try to rebel against this new reality, and then a long, slow grieving process as you attempt to accept a world that looks very different from the way you wish it did.

Former client (C)

The experience of ecological grief[6] is being increasingly understood as being a tandem emotion of climate anxiety. This might be a direct experience of grief if a loved one, or your home, is lost from a natural disaster; equally, as a more indirect, almost premeditated grief that occurs when someone feels the deep pain of what future world their child will inhabit or perhaps the loss of species as rainforests are burned.

We would be remiss not to mention Swiss psychiatrist Elisabeth Kübler-Ross's five-stage grief model,[7] which was inspired by her work to take a more honest approach with terminally ill patients. She describes a process that most people who have ever dealt with traditional grief will be familiar with. This process includes denial, anger, bargaining, depression and acceptance. We can simultaneously appreciate that this book revolutionized the culture, commonplace in Western society in the post-Second World War era, to avoid the negative emotions associated with death by effectively gaslighting or washing over them, but also accept that the criticism it received for its overly linear understanding of the grief process. We can benefit by acknowledging that the experience of grief is difficult to model, and although the framework can be useful, there is certainly no 'right way to grieve'. It's a highly personal experience and, especially in the context of climate change, very fluid.

Psychotherapist and climate change researcher Rosemary

Randall identifies how Kübler-Ross's model fails to fully capture the experience of climate-related grief because it details a process of accepting the inevitable end to a life (an individual's or that of a meaningful relationship). It is final and absolute and there is little room in this context to engage hope or possibility. Randall observes:

> In contrast, when facing climate change we still have much to hope for and much to play for. The changes and adjustments we make also need to last across substantial periods of time. Sustaining our creativity and resourcefulness is essential. We have the chance to remake our lives.[8]

Randall also describes a typology of loss:

- **Absolute loss**, which is distinctive in its finality in relation to the climate. This could involve the extinction of a species or a forest burning to the ground, or the death of a loved one in a wildfire.

- **Chosen loss**, which involves us voluntarily deciding to forego something that we cherish for the greater good, or because our personal values conflict with the pleasure and meaning it may bring. For instance, choosing not to have children for ecological reasons.

- **Transitional loss**, which describes the grief associated with leaving one stage for another – for example, becoming an adult and leaving childhood behind. When we think about this in terms of climate change, it can become less clear as we aren't sure what the future may hold. It could be a utopia of both ecological and systemic repair (our 'Symbiocene') or it could spell the end of human life on earth as we know it.

- **Anticipatory loss**, which is a bit like the pre-trauma we sometimes envision when we think about climate change. We pre-mourn what may be lost. We begin to process our grief in order to prepare.[9]

When we understand more clearly the kind of grief we are experiencing, we can lean into the acceptability of this being an incredibly heavy emotion for us to bear, and rather than berating ourselves or silencing grief, we acknowledge that we are experiencing it because of the fear of losing what we love and care about. When we sit with this, we can encourage self-compassion, engage others for support and then feel able to reinvest these feelings in climate-positive action. Or, as Randall says, 'When loss remains unspoken, neither grieved nor worked through, then change and adjustment cannot follow.'[10]

Climate anger

VOICES

Mourning – a deep depressing grief for the planet, something which is so much bigger than myself, something so beautiful and spiritual that we as humans have no right to interfere with. Energy sapping, heartbroken at what we may lose, angry at those who are creating such devastation and frustration at those who cannot see what is happening and how rapidly we must change.

Former client (S)

Anger and anxiety share many similarities, though they feel very different. At a neurochemical level, they are similar. Indeed, the FFF processes that occur in both states are much alike. Have you ever noticed your heart beating harder and faster when angry? Or, likewise, found your jaw or fists clenched? Hot? Shaking? In a broad sense, you can almost equate anger to the 'fight' response and anxiety to the 'flight'. In other ways, too, psychologists often link both anger and anxiety to a loss of control. We tend to experience either anxiety or anger (or both) in response to situations in which we feel powerless, or where there is a degree of uncertainty as to the outcome or how we should respond. We can fluctuate between these two emotions readily, and so often our response to stressful events is non-linear, involving multiple, sometimes conflicting or competing, emotions.

When it comes to climate-related anger, it can be easy to see how it can be triggered. Climate change constitutes a high-magnitude issue, with a low level of personal control (at least, in that we are limited in terms of how much influence we can have over the actions of other people). In psychological terms, this creates an ideal scenario for anger to occur.

We may feel anger about how others are responding to the climate emergency. It's important to be mindful that people process difficulty in different ways. It's more useful for us and the planet to channel the frustration we feel towards others into motivation that drives climate actions. This is not to say that we shouldn't engage in conversations and promote acceptance and action whenever possible – we absolutely should – but we need to assess the value of our time and energy. We need to take stock of where we can have the greatest impact. Essentially, we can 'get mad' or 'get mad and get to work', but we will find strength and resilience in the latter. Often when we do this, our anger naturally evolves into more positive and connective emotions.

Climate depression

Solastalgia describes both grief and sorrow, and, as such, these two are often interwoven in people's climate emotions. Depression is strange in many ways and yet in others makes perfect sense. Perhaps climate depression (again, we are not talking necessarily about a 'clinical depression' here) is as nature intends. Is it our way of connecting with the changing planet? Of reminding ourselves of the inescapable interconnectedness we have with the natural world? As Per Espen Stoknes considers: 'Depression is a state of mind that brings...a sense of limits and humility. Through it, we may discover that we're not on top of the world. Rather than being masters in control, depression teaches that we are fully dependent.'[11]

The ways we experience depression, when tasked with describing it, can vary so much that one would be forgiven for thinking we are talking about very different ailments indeed. For some, apathy. For others, anger, disconnection or even physical pain. Depression often brings with it a set of cognitions that we sometimes call 'the voice of depression'. This voice may reinforce feelings of helplessness and lack of hope for the future. It can also bring with it a more dangerous set of thoughts, such as a temporary belief that 'it would be better if I were not here at all'. Therefore, depression and suicidal thinking can, sadly, go hand in hand. Although climate depression is not the same as clinical depression, there is the potential for overlap, and in its extreme form, thankfully rarely, suicidal thinking can be held in positive regard as a tragic solution to carbon neutrality. It is crucial, therefore, that if you, or anybody you know, is experiencing thoughts such as these, to seek professional support (which can include going to your local accident and emergency department).

Climate guilt

VOICES

Really try to understand that while there is a lot you can do to help tackle climate change, this mess is not your fault...and that you need to take care of yourself and act within your limits.

Former client (K)

In the wise words of Megan's beloved Italian Catholic Godmother, 'Oh, just let the guilt go! It doesn't help you at all.' Rather, look to guilt with curiosity. Is it revealing an inconsistency with your values? Is there perhaps a utility to this uncomfortable emotion, after all?

We can think of guilt as a flashlight, illuminating situations in which we may have suppressed or were incongruent with our own moral behavioural code. For example, if you were unfaithful to a partner and you experience guilt about this, it highlights the values of trust and honesty that were contradicted by your own behaviour. By examining those times in our lives when we feel either concordant or discordant with our values, we can begin to tease out these feelings of guilt and 'thank them' for what they can reveal and how we can better show up next time. If you feel as if you did something wrong, stand up and own it. Sometimes this means apologizing or making amends, but creating self-inflicted discomfort whose only purpose is to remind us of our mistakes will impede us from moving forward into a more positive and productive space. *Just let it go.*

Guilt can also be useful in the sense that it can signal a predisposition to some perfectionist tendencies. As we acknowledge, perfect gets nothing done and can stand in the way of us working towards and achieving our objectives. In terms of climate guilt, this goes hand in hand with accepting that all the blame does not fall to us. What use is it to look back to those few times you left your re-usable carrier bag by the front door? Or the fact that you are part of a system with a disproportionately large carbon footprint? We are not advocating for complacency or permission to accept the inherent broken and unjust systems in which we live. In fact, a great way of channelling climate guilt is by engaging more in trying to affect wider policy change, which is the level at which the necessary collective change needs to occur. As Leah Stokes highlights: 'The goal is not self-purification. The goal is institutional and political change.'[12] We do not have to feel bad about being imperfect.

Positive climate emotions

VOICES

I really, really enjoy the activism. I do. Because my activism doesn't stem from a place of anger or fear any more. It used to, but it's developed into a place of love – a love for the people and love for the planet.

— *Mitzi Jonelle Tan, climate justice activist from the Philippines*

The umbrella of climate emotions is large and certainly, without question, includes positive ones. These can be confusing because we may feel bad for feeling okay. However, our clients and collaborators often describe the positive side-effects of climate engagement. These include an immense feeling of connection towards people and nature, feelings of pride in their work, motivation and commitment, with a belief that they are powerful. Going beyond feeling connected, we've been inspired by the stories of activists and climate workers who describe the love that they feel for humanity. A deep love that promotes their wellbeing and ability to stay abundant and resilient in their work, and likewise protects them when they feel the negative emotions surrounding climate change.

Saying 'thank you' to your emotions

Chances are, if you have chosen to read this book, you will, at some stage, have experienced climate emotions that have felt unhelpful, perhaps overpowering or inescapable. What we are about to suggest may therefore seem an odd exercise, but many people have found it useful in order to establish a healthy relationship with their emotions, which, let's face it, will always be a part of our lives in one form or another. So, for this exercise, simply note below three things that your climate emotions have given you, for which you can be grateful. It may be a time when anxiety or anger served as an impetus for some sort of change, or a time when grief or sorrow connected you, at an emotional level, to the natural world. Likewise, have you formed meaningful friendships and communities after connecting through your shared emotional experiences of climate change? See if you can bring to mind specific examples from your own life, and record them below:

Exercise: Dear climate emotions, thank you for...

1. . . . .
. . . .
. . . .

2. . . . .
. . . .
. . . .

3. . . . .
. . . .
. . . .

How we communicate with ourselves when we experience strong emotions is incredibly important. By taking the time to accept that doing so is a normal and natural part of who we are, and acknowledging that they are essentially trying to keep you (and the world) safe, happy and connected, you can approach them, and therefore yourself, with a greater degree of kindness.

Chapter 5

Defences, Biases and Climate Behaviour

The psychological defences

In the words of Jim Butcher:

> [T]he human mind isn't a terribly logical or consistent place. Most people, given the choice to face a hideous or terrifying truth or to conveniently avoid it, choose the convenience and peace of normality. That doesn't make them strong or weak people, or good or bad people. It just makes them people.[1]

To experience some form of psychological defence (our unconscious protective mechanisms against psychological vulnerability or distress) in the face of threat is universally human. Quite simply, if all of our defences were to be suddenly stripped away, the world would be a truly terrifying place indeed.

Our defences help us cope with the multiple demands, triggers, disappointments and losses in life. So often, when we hear the stories told by the people we work with, it feels like holding a mirror up to ourselves and our own ways of responding to the world. As psychologists, we are no different, and we have plenty of our own defences, too (many of which went into the making of this book)! Constructs such as denial, avoidance and projection are truly nestled in the popular understanding of human psychology. They are common parlance. They are also great examples of our psychological defences at play. However, they perhaps don't get the credit they deserve because, in certain ways and in sufficient but not excessive quantities, they allow us the peace to enjoy our lives.

Our defences can, of course, be problematic, too. But how do they become activated? How can we learn, and then unlearn, their ways? Our defences are essentially the ways in which we keep ourselves safe, by neutralizing the psychological threat. It was Freud (first Sigmund and, later, his daughter Anna) who first described and charted the 'ego defences'. However, they are not limited to Freudian psychodynamic theory, nor the psychoanalytic method. In more recent incarnations of therapy, the defences are considered but called something else. In the case of CBT, we might work to better understand our 'core beliefs' about the world, and the 'safety behaviours' we use to protect ourselves psychologically. We might also work to understand 'unhelpful coping mechanisms' that sound familiar; we avoid, procrastinate, distance and intellectualize. Even apathy (or indifference) to climate change can be viewed as a deeply manifested protective defence mechanism against its harsh realities, being instead, in the words of psychologist Renée Lertzman, part of a kind of 'environmental melancholia'.[2]

The Five Ds

Per Espen Stoknes presents perhaps the most succinct explanation of the characteristic ways in which our defences become activated in response to the climate crisis. Or, specifically, in response to some form of climate information or message. These defences, he suggests, form the psychological barriers to climate action. His five defences (or Five Ds) chart a landscape of unconscious psychological processes, whose intention is to undoubtedly preserve our wellbeing, and do so quite successfully in the short term. However, in the longer term, they can hinder our sense of self-efficacy and prevent us from engaging with the problem.

So take a moment to consider these defences. As Stoknes suggests, notice how the concentric circles around the person move from the outermost (Distance) to innermost (iDentity) defences, where iDentity describes the relevance of the fundamental ways in which we view ourselves. Allow yourself to bring these defences more into your conscious awareness. By doing so, it allows you to take greater agency over your responses to climate information, and in doing so, guide yourself towards climate action. Or as Carl Jung reminds us, 'Until you make the unconscious conscious, it will direct your life and you will call it fate'.[3]

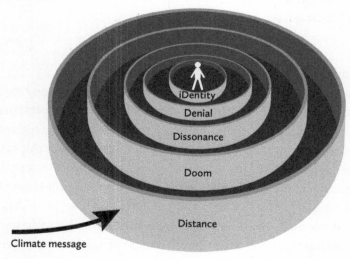

Figure 5.1 Per Espen Stoknes: The Five Ds (Used with permission)

Distance

When confronted with climate information, we want it to feel far away, in both space and time. We often try to distance the issue, as something affecting faraway places, in a time far from now. Historically, this effect has been mirrored by the way stories have been presented. Additionally (although as time progresses and climate events become more evident around the world), these stories have, for many of us, quite simply occurred in places that are indeed far away from our daily lives. Seeing glacial melt, for those of us in temperate or subtropic regions, doesn't always conjure an immediacy, at least not as a threat to our immediate environment, our family or community. The language used has often perpetuated this sense of distance. Our perception of risk is rarely affected by numbers, however stark they are, but instead by stories that feel close to us. Not 'then' but 'now'. Not 'there' but 'here'. For this reason, it has been argued, the climate message, often involving overly statistical representations of the risk, and trying to encourage us to care deeply about a far-off, future scenario, has allowed us to maintain our defence of Distancing by creating an 'intellectual' or hypothetical risk.

Doom

If, when the problem is presented as overly far-off, we engage in Distancing, what happens when the problem is instead presented as immediate, catastrophic, here and now? The scientific community has been at times, understandably, at a loss as to why this message has not resulted in a greater response. In essence, the opposite can happen. We can become over-exposed and then immune to bad news, or, at least, a kind of doom fatigue can set in, which, if accompanied by the lack of a clear, practical and achievable set of personal actions, can coalesce with a sense of helplessness to push us further away from engagement.

Terrifying headlines that evoke the scale of the change necessary can counterintuitively serve to demotivate us, although this isn't always the case, and not for everybody. Jem Bendell's 2018 paper 'Deep adaptation: A map for navigating climate tragedy'[4] struck a particular chord, and created an almost unprecedented amount of noise for such a publication, being downloaded hundreds of thousands of times, and translated into a dozen languages. It was even suggested to have provided inspiration for the Extinction Rebellion movement.[5] However, many notable climate scientists have refuted Bendell's message as overly fatalistic and doom-laden,[6] while others agree with the overall sentiment. It is not our purpose to either agree with or refute Bendell's claims. The point is that this approach isn't going to resonate with everybody, and who knows the degree to which it activated the psychological defences of the scientific community's response. It certainly risks igniting our doom fatigue. As Stoknes describes: 'We've heard that "the end is nigh" so many times, it no longer really registers.'[7] If we understand the message to be 'we are screwed and there's little we can do about it', where's the purpose in trying?

Dissonance

The space between the self and the ideal, in whatever form that may take, constitutes dissonance. Or, more specifically, cognitive dissonance occurs when there is some degree of tension between our own values, beliefs and behaviours. In his theory of cognitive dissonance, psychologist Leon Festinger suggested that people crave internal coherence. When external messages contradict how we think and what we do, this creates an uncomfortable psycho-logical dilemma, which is remedied by attempts to restore this coherence, often by dismissing the dilemma-inducing message. Or as Festinger describes: 'A man with a conviction is a hard man to change. Tell him you disagree and he turns away. Show him facts or figures and he questions your sources. Appeal to logic and he fails to see your point.'[8] Herein lies the issue. We should not be 'convicted' in our own beliefs and attitudes. We should remain open to different possibilities. More often than not, conviction emerges not from a careful weighing up of the evidence for and against an argument, but instead retreating to a sense of inner coherence, to prevent the activation of discomfort arising from some form of cognitive dissonance.

We also adopt other ways of maintaining inner coherence. We may downplay the negative potential of our own carbon impact – 'Our neighbours have two cars and we only have one' – or we downplay the strength of the evidence that what we are doing harms the planet, or likewise the benefit of the dissonant behaviour – 'Did you know that avocados actually wreak environmental havoc?'[9] The latter may be true, of course; the avocado industry has grown exponentially beyond what is arguably scalable and sustainable, but does that necessarily justify eating red meat? Dissonance would dictate that these could be used as 'excuses' to justify our own non-environmental behaviours. Two wrongs don't make a right.

Denial

Our minds allow us the ability to both 'know' and 'not know' at the same time, to protect ourselves against some sort of trauma or dilemma. This is denial.

Stoknes is right to identify the negative connotations associated with the word 'denial', particularly in the climate arena. Climate change 'deniers' and 'denialism' have become politically weaponized terms, often used to describe a systematic denial of the scientific evidence for some sort of personal, financial or political gain. There are plenty of weaponized terms used by these groups too, of course, such as 'hoax' and 'climategate',[10] to describe the so-called conspiracy of climate science. Put simply, people often reject information that contradicts their own beliefs or world view – so-called motivated cognition.[11]

Confusingly, too, denial can also be used to describe awareness and acceptance of the evidence for climate change, but without the associated level of personal action.

Remember, however, denial in some ways is protective. Or, at least, it comes from a place of self-protection. It is not ignorance or arrogance.

iDentity

Now we journey into the innermost circle, that which is closest to our core view of ourselves and our place in the world. Our identity, even politically, enters the arena and can have huge influences not only on our interpretation of science but on our climate-related behaviour. Climate change has long been associated with left-wing politics, but why? Well, in many ways, the liberal political viewpoint lends itself to the idea of collective action for the greater good, even if that means, for example, carbon taxes, or other forms of top-down regulation, whereas those of more conservative or libertarian political persuasions would traditionally reject the idea of state-imposed regulations and limitations on individual liberties.

So, when presented with climate messages, particularly involving the actions required, these can sometimes find themselves battling against a person's very sense of identity, and losing. Or, as Stoknes puts it: 'The messages crash against the wall of the self.'[12]

My Five Ds

So, to continue the journey inwards, what do these look like for you? In a Jungian way, make the unconscious conscious, and by doing so, empower yourself with the knowledge of your, often clandestine, psychological operators. Manning the controls without your knowledge. What are your Five Ds? What sort of information activates them, and what clues may you have that they have been activated? What behaviours, emotions or thoughts might you notice?

My Distancing defence:

. . . .

. . . .

. . . .

My Doom defence:

. . . .
. . . .
. . . .

My Dissonance defence:

. . . .
. . . .
. . . .

My Denial defence:

. . . .
. . . .
. . . .

My iDentity defence:

. . . .
. . . .
. . . .

Cognitive biases

Humans think imperfectly. However, this imperfection is itself crucial to our survival. In his book *Thinking, Fast and Slow*[13] (which followed decades of research into human decision-making), eminent psychologist Daniel Kahneman describes two distinct cognitive systems. The first (System 1) is designed to make rapid decisions when time is of the essence. We must make heuristic use of the evidence we have in front of us to quickly decide upon our response. This system is particularly prone to errors. Our second (System 2) is slower and more deliberate. We take the time to analyze the data, weigh the pros and cons, consider multiple perspectives. This system is less prone to error but it takes a certain amount of time and psychological bandwidth.

Along with these two systems, there are a number of predictable biases in our thinking. These biases pervade our daily lives in sometimes productive and other times unproductive ways. Shall we meet some of them?

Confirmation bias

Much has been made, with good reason, about the social media 'echo chamber'. Users find their own attitudes and beliefs being reflected back from the walls of social media platforms, thus strengthening their original standpoint. We are much more likely to welcome, and engage with, news stories, posts and advertisements that echo our own sentiment than contradict it. Architects of social media platforms are, of course, aware of this and create algorithms that make us engage more, stay for longer and produce more of our own content. This is, if you like, our 21st-century digital confirmation bias, but we can hardly blame social media. We have a long history of rejecting evidence because it doesn't fit with our own beliefs, at times with widespread and long-lasting consequences.

With relevance to climate, though, confirmation bias rears its head everywhere. Of course, it enters the conversation in terms of climate change denialism; however, we are all susceptible to more readily accepting evidence that supports our pre-existing mental models. This is often activated in the face of short-term weather events, particularly brief spells of unusually hot or cold weather, as evidence for (or against) our pre-existing beliefs around climate change. This is problematic, of course, when a climate change sceptic experiences unseasonably cooler-than-average temperatures and sees this as evidence that global warming is a hoax, thus confirming or strengthening their beliefs. Likewise, however, the reverse can happen, where unseasonably warm weather may accelerate a climate-concerned person's belief that the changes are happening more quickly than the evidence suggests, leading to potentially unhelpful amounts of anxiety and panic. The answer to both, as the scientific community repeatedly highlights, is to focus on longer-term environmental changes as more robust evidence of global warming, rather than short-term seasonal effects. However, we can see how short-term weather events can activate our pre-existing confirmation bias in either direction.[14]

Status quo bias

Simply put, the status quo bias describes our preference for what already is. We return to this later in the book, where we look at ways in which we can all use the status quo bias to the planet's advantage, by manipulating the 'default options' presented to people, knowing they are most likely to choose the option that best resembles how things currently are. There's a joke that we psychologists love (and probably we are the only ones who laugh quite so heartily when we hear it). 'How many psychologists does it take to change a lightbulb?' 'Only one, but the lightbulb has to *want* to change' (Megan rolls her eyes).

You see, we are hardwired to automatically favour non-change. Or, at least, to experience aversion to the idea of change, even when we can rationally understand the change as being, all considered, a good thing. Provided the status quo is 'good enough' in our minds, we are likely to make decisions (fast, automatic 'System 1' decisions) that perpetuate the current state of affairs. This is arguably because decision-making, particularly involving a new scenario, carries a degree of risk. Also, losses loom larger than gains. We tend to be risk-averse in our decision-making, not least because we find a loss more deflating than an equal gain is enjoyable. So System 1, quite without our realizing, quietly asks, 'Why take the risk?'

Needless to say, when it comes to a problem that fundamentally requires change, the status quo bias is less than ideal.

The 'thinking trap' biases

If the above biases relate to the predictable ways that our thinking occurs on a daily basis, in a fairly innocuous manner, the next section shines a light on the ways in which strong emotions such as anxiety, anger or helplessness can affect our thinking, by creating 'cognitive distortions' (or 'thinking traps'). We will have all experienced one or more (if not every single one) of these at some point in our lives. They tend to become 'activated' when we are feeling stressed, anxious, angry or sad. When we feel this way, our 'high order' cognitive processes, which allow us to hold multiple perspectives, decision-make, plan and rationalize, become affected, just when we need them most. Put simply, it becomes much more difficult to think in a slow, balanced and reasonable way when we are experiencing strong emotions. Instead, we find ourselves slipping into one or more of the thinking traps.

The thinking traps are problematic because they actually reinforce negative emotions and unhelpful patterns of behaviour. For example, if I had claustrophobia, I might *feel* scared entering an elevator, and the thinking trap of 'emotional reasoning' (see below) would become activated, I would then judge reality based on the emotion I was experiencing, rather than the actual level or risk: 'I *feel* scared (unhelpful emotion), therefore I must be in danger.' I feel as if the elevator is going to plummet, therefore I convince myself this is a likely outcome. This might lead to me avoiding elevators (the unhelpful behaviour), which would perpetuate not only the fear but also the belief that elevators are dangerous.

Climate change is not the elevator, of course. The threat is more real. However, these thinking traps can still occur in the context of climate change and similar unhelpful cycles can emerge. These can affect not only our wellbeing but also how we engage with (or avoid) the issue of climate change itself.

Here are examples of some of the main thinking traps, and how they may be activated in relation to climate change. As you read through these, ask yourself whether you have experienced any or all of these, and, if so, what situations trigger them for you? Further in the chapter, you can record your most frequent or problematic thinking traps, and then learn how to challenge them later in the book.

Emotional reasoning

For example: 'I feel powerless in the fight against climate change; therefore there is nothing I can do'

Emotional reasoning is the process of assuming truth based on how we feel. Our emotions are exceptionally powerful at shaping our interpretation of a situation, particularly its potential outcome. If we *feel* scared, we might predict danger. If we *feel* hopeless, we may predict a negative outcome or perceive ourselves as not being able to affect change. If we *feel* guilty, we may believe that we *are* doing all the wrong things. However, feeling scared in itself is not evidence that we are actually in danger. It is only evidence that we feel scared. Of course, when it comes to climate change, there are plenty of negative predictions already, many based on sound scientific evidence. However, if we *feel* helpless or hopeless about the future, it may undermine our sense that there are things we can do, which would lead to demotivation.

'All or nothing' thinking

For example: 'If I don't work tirelessly and constantly to fight climate change, I've failed'

The 'all or nothing' thinking trap (often called dichotomous reasoning, or 'black and white' thinking) prevents our ability to hold a balanced view or to appreciate the shades of grey. We instead view a situation, or our involvement with it, in extreme terms. One example regarding climate change may be to experience the thought that 'I must change all aspects of my behaviour right now or else there's no point in trying'. This can often be reinforced, for example, if we attempt too much too soon, and experience a failure to enact a behavioural change, and we then give up altogether, thus 'reinforcing' (at least in our minds) that there was no point in trying.

The mental filter

For example: 'Nothing I am doing is making a difference'

The mental filter can almost be seen as confirmation bias in action. Our attention can only ever focus on a particular selection of all available information at any given time. Of course, when we are anxious, we selectively attend to threat stimuli. When we feel down, or self-critical, we tend to selectively attend to 'evidence' that may confirm the negative feelings. As a result, the mental filter ends up perpetuating the echo chamber in our minds, and the unhelpful beliefs go unchallenged. With a problem as big as climate change, we can be quick to dismiss (or fail to attend to) the small successes we have along the way. We may disregard our positive climate actions, instead focusing solely on the changes we haven't yet made, or holding ourselves overly accountable for the attempted changes that aren't going so well.

Magnification and minimization

For example: 'My successes are so trivial'

Magnification is the process of attributing more weight to our perceived failures, and minimization is the equal and opposite process of devaluing our achievements. It is closely related to the mental filter. So you may find yourself not ignoring your successes completely, but instead giving them proportionally less importance than your perceived climate 'failures', or, likewise, being quick to dismiss (or minimize) the impact of the work you are already doing.

Personalization

For example: 'This is all my fault'

The personalization trap occurs when we attribute blame to ourselves for things outside our control. This creates a complicated conundrum in climate change, given the human contribution to the problem. In this context, it's best to say that the personalization trap causes us to *over-attribute* blame to ourselves or hold ourselves overly accountable as individuals. Although accountability can be a good thing, being overly self-punishing can actually act as a demotivator, as well as impact our wellbeing.

The first step in overcoming thinking traps is to identify them. Once you have an idea of the way your thinking can be negatively affected in relation to climate change, you can become empowered to challenge them. This in turn can create a virtuous cycle of positive thinking, which can therefore motivate and inspire you to enact change. Using the space below, try to record some examples of how your thinking may at times be affected, specifically relating to climate change.

Exercise: Understanding my own eco-thinking traps

Situation Where was I? What was I doing?	Negative thought What thought entered my mind at that time?	Thinking trap Which (one or more) of the traps became activated?

Doomscrolling: digital rumination

The term 'doomscrolling', which entered into the public consciousness as a result of the multiple global existential events of 2020, refers to perpetually scrolling through the internet and attuning to negative information, and became one of the key issues that we identified as a trigger (and perpetuating factor) for climate anxiety. Doomscrolling most commonly occurs online, but some people will consume newspapers or television in the same way. The process happens because people see information but lack concrete solutions about how they should respond. They then attempt to seek out more information, essentially looking for answers that will resolve the threat. This can spiral, leading to hours spent looking online and feeling all the worse for it.

Doomscrolling thrives off that old part of our brain (the limbic system) that was used to alert us to those cave-person dangers but doesn't quite compute when it comes to more modern existential threats, as well as the constant pull for our attention from multiple sources. This other component is the emotional yo-yoing that we tend to do online. We might casually open our phone and click on an article predicting, or reporting on, current climate disasters. We feel our anxiety rising. Our algorithm is clever: we are fed more and more negative stories about the crisis. We feel our throats tightening and our stomachs turn. We are interrupted by a video of cats falling from things, then back to flash floods in Afghanistan. A WhatsApp message from Mum pops up while you're trying to figure out what you can actually do to help the planet. Oh, and this is all before you have arrived at work and you also missed your bus. Emotionally exhausted yet?

When we are constantly searching for this threat and bouncing between emotionally provocative content, this can lead to a sort of hypervigilance that is hugely mentally taxing. 'Smobies' (smartphone zombies), a phrase coined by *Radical Attention* author, Julia Bell, describes how: 'Over time, trained by the software, the user...

oscillates between anxiety and the alleviation of that anxiety, over and over, and bouncing between those positions makes it impossible to think.'[15] When we are in information consumption overdrive, we don't take the necessary steps that we normally would if the threat was more tangible and we would know how to overcome it. Without exercising the action, we retain the emotional discomfort; therefore, our action is to seek further information, and the process repeats. When we doomscroll, it's also difficult to disengage because it produces a compounding effect that encourages us to keep seeking information and confirmation about the danger and less frequently to search for potential positive solutions. This may also undermine our sense of personal agency in the problem and can leave us feeling depleted and defeated.

The design of social media platforms also has a part to play. Their 'endless' nature means that the reader has to be particularly proactive and 'switch off'. Likewise, algorithms perpetuate the content people engage most with, creating quite readily a doom loop of frightening information. Without actively switching off, the potential consumption of messages can be endless.

Unsurprisingly, doomscrolling became a widespread problem in 2020, given the multiple existential crises we faced, and the amount of time (in COVID-19 lockdowns across the world) that we were afforded with our phones, laptops and televisions. These contributing factors created a perfect storm (if you'll forgive the pun). So much so that *Washington Post* writer Sunny Fitzgerald described doomscrolling in 2020 rather brilliantly as 'the tangled relationship of human survival instincts and technological design amplified by the pandemic'.[16]

VOICES

I used to read every environmental news article published. I have realized that these just feed anxiety and very rarely tell me any thing significantly new and so I find that avoiding these stories or significantly reducing my engagement with them has helped my climate anxiety.

Former client (S)

'Troll scrolling'

Another issue that can trigger anxiety or anger, which again stems from the time we spend on digital platforms, is the increasing sense of polarity in online conversations. This can mean conversations escalate quickly into arguments. From mask-wearing to our food proclivities, there is the sense that someone out there is aching to jump into the comments and tell us why we're wrong. Online, people seem to feel more willing to express their, often profanity-laden, opinions. What happens when we engage with this? Have we found ourselves reading the comments of a post and, before we know it, we are 146 comments deep and feeling our blood boil? We might describe this as 'troll scrolling'. Perhaps we too have commented, so we check repeatedly for how it has landed. Or, instead, maybe we spend the next few hours thinking about how to craft our own response. When we do this, we need to stop and ask ourselves whom, and to what purpose, does this interaction serve? How much mental energy is this worth, and am I likely to shift their beliefs?

The field of social psychology has long taught us that, in most cases, when two people scream opposing arguments at one another, the result is a reinforcement or even further entrenchment of their own views. This can create further frustration: 'Why aren't they listening to me?!' We also know that social media platforms, while providing an unprecedented opportunity for connection and social affiliation on a global scale, are also spaces where people experience a greater sense of political disagreement when compared with face-to-face interactions.[17]

There is a slew of potential emotional landmines planted in the comments sections of these feeds that may trigger us. They are inherently platforms that encourage dialogue and interaction. When well informed, factual and productive, they can positively disseminate knowledge and encourage engagement and action. But when we are feeling the weight of climate change, we need to decide in advance our level of engagement in the shared beliefs of others. If you have gained knowledgeable facts or credible evidence to counterbalance an argument that has been offered, by all means engage. If you are feeling robust in your wellbeing and want to be a part of the dialogue, engage. But we encourage you to take your resilience pulse first.

Reading into comments, how many times have you engaged in social media or a WhatsApp chat and swayed the opinion of someone who is a staunch critic of the statement you agree with? If this happens a lot for you, then carry on because your online communication is clearly effective – please teach us all! If climate emotions are ringing loudly for you at a given time, we encourage introducing a self-imposed 'like and scroll on' policy. If you see a post that you like and agree with, which is true (and by true, this can also mean a statement or image that speaks to you), by simply 'liking' or 'liking and sharing', you have participated in an online action that can end there. In a similar way to how we can vote with our wallets by buying green, local, Fairtrade products and banking ethically, we can spend our *online* capital by supporting and promoting climate-conscious, accurate information. It's essential that we determine that the facts we promote are true, but 'liking and sharing' this content is a climate action that has an impact.

Megan recalled the Fridays for Future events that she attended in Oxford. For a while, whenever she found herself thinking back to these protests, something both annoying and illuminating surfaced for her. Out of all the people there – the activists and scientists, academics and kids, the parents, lawyers and teachers, farmers and health service workers – when she thought back to all of those standing up for the climate, there was always an intrusive face that popped up to say 'hello'. One man would invariably show up and make a performance of being in a conversation on his phone (who knows if there was even anyone else on the line?). He would deliver a running commentary on the phone about 'the lefties', 'the communists', 'the vegans', the 'socialist environmentalist idiots' (with plenty of disparaging language to accompany this), along with vague threats of what he would like to do about them. This was just one person. But this annoyed her because out of the sea of hope and resistance and action, this face popped up to taunt the efforts of all those who showed up to say they want a better planet.

Who cares, really? But this felt bothersome to her because *one* man robbed a little of her power that day. That man's face signifies the face that we are sometimes consumed by if we don't weigh in the power of the engaged, and our own power. If we forget that others are pitted against their own defences, narratives and experiences. If we allow ourselves to invest emotionally in them, they detract from the focus – caring for the planet. For all those (and there are fewer and fewer) that say 'No', we can look around us, connect and say, 'We will anyway'. When we take the bait, it only depletes us of our energy that would be better spent on our climate work. This served as an important reminder to step back from the drama.

Does this online behaviour sound familiar? Don't worry if so: you are in good company and overcoming this can be pretty straightforward as long as we exercise everyone's favourite thing – self-restraint. But take a moment now to think about your doomscrolling habit. It is very much a habit and, as such, can be predictable and predicted. Understanding how, when and where you doomscroll allows you to take the upper hand and create a bespoke digital diet that works for you.

Exercise: Charting my doomscrolling

Where and when am I most likely to doomscroll (e.g. at night, in the morning)?

. . . .

. . . .

. . . .

What are my doomscrolling triggers (certain news stories, social media channels)?

. . . .

. . . .

. . . .

How do I feel after doomscrolling?

. . . .

. . . .

. . . .

What protective post-doomscrolling measures can I take to regain a sense of power? Can I limit the amount of time I research online? What can I do afterwards to help 'bookend', and restore my wellbeing?

. . . .

. . . .

. . . .

Chapter 6

Communication About Climate Change and Why It Makes a Difference

Climate communication has tended to fall down in the past because the scientific messages haven't always translated in a way that engages a positive psychological or behavioural response. We've traditionally received climate messages that delivered the facts about climate change in a way that was, well, pretty miserable. It was all the things that had gone, and would go, wrong. An uninhabitable planet, a fireball, mass extinction and, for some, the most horrendous notion that they would need to change their comfortable and convenient lives to abate it (triggering all kinds of dissonance). 'Make my burger a double, and could you hurry? My flight is about to leave.' So, it is no wonder that we, and many others, wanted to dismiss climate information, or at least diminish it. Perhaps to its own PR detriment, it's been labelled *An Inconvenient Truth* for a reason.[1]

When we think about climate communication, it might elicit feelings of frustration that the message is not getting through. 'Why am I the only one who seems to get it? How are my friends and family just not bothered about the climate crisis?' This can feel isolating or infuriating. We may experience times when we are presented with a fact that is upsetting and we may feel as though we are just left hanging, like being told by someone, 'You are on fire' and then they just walk away: 'What do I do with that?' There isn't much containment. Or we might feel guilty: 'Why didn't I listen sooner? Why didn't I take more exceptional measures years ago?' (a little personalization). Likewise, 'I just can't hear this today.' When we understand a bit more about the psychology of climate communication, we can learn to feel more compassion, because what's at play when we receive messages about global warming are our many defences and biases doing their best to save us the misery that total exposure to climate information might bring.

When we understand that much of this turmoil couples with the issue that there is a general problem coming from science and media communication systems, and, more alarmingly, politicized and polarizing messages that distort our relationship to climate information, it becomes easier to practise empathy, awareness and compassion towards ourselves and those who have trouble taking on this information. This allows us to be more open when we receive these messages and pragmatic with the information we get. Likewise, we can become more strategic and conscious when we discuss global warming with others.

There are amazing resources available to teach people the most effective ways of communicating about climate change. The Yale Center for Climate Communication[2] runs fantastic online courses, which are accessible to everybody. Likewise, national efforts such as Britain Talks Climate[3] address directly the question of *how* to talk about climate change, regardless of the platform or nature of the interaction – at work, at home, with friends. We just want everyone talking about this.

Researchers, governments and NGOs have tended to run into difficulties when constructively communicating about climate change. Surely, if people hear information that is grounded in robust science, and is also stark, alarming and urgent, along with information on how they can respond to these challenges, then they will accept these facts and take actions to avert the problem. This must be a win–win–win? From an evolutionary psychology perspective, communication about climate change can seem complicated. In a straightforward situation, if somebody with credible knowledge warns you of danger, you take action. For example, if a policeman tells you that there is a criminal on the loose in your neighbourhood, and that you need to go into your house and lock the door, you probably would. But when experts warn us of *potential* or *distant* danger, our basic instincts are less able, or less inclined, to attune to this threat and to take action. Examples are useful so let's have another: an anti-smoking campaign. There is a doctor in a white coat, shaking his head and pontificating to a 16-year-old about the dangers of smoking on one's health, telling them that they should really quit (oh yeah, and all of this kid's friends smoke, which doesn't help). This will resonate less than when a doctor says to a 55-year-old cancer patient that it is time to knock it off if he ever wants to meet his grandkids. What we have is the 'there/then versus the here/now' conundrum (a problem of distance).

Climate change communicator Susan Joy Hassol notes that 'Scientists' training teaches them to use jargon, and include all the details and caveats right up front. Never repeat what is already known, and let the data speak for themselves – there's no need for stories.' We can see that this is in direct opposition to what most people need, in order to attune to climate information. She goes on to identify the essential elements of climate communication for the public: 'Simple clear messages in plain language, repeating them often and communicating using stories because they are sticky and memorable.'[4]

Even the *language* we hear in climate communication can cause confusion. Is it 'global warming' or 'climate change'? 'Crisis' or 'emergency'? Should we call it 'natural gas' (natural must be positive) or instead might it better help shift public attitudes to start saying 'methane'?[5]

Although Hassol identifies that framing, language and accessibility (in terms of information that can be understood by the given audience) are some of the ways that the public can be *unintentionally* misled about climate facts, she also importantly highlights that communication can be *deliberately* weaponized. This is done to sow the seeds of disinformation in order to exploit the understanding of global warming, carried out by motivated political and economic groups, in order to maintain the status quo for the fossil fuel and other industries. This polarization for political gain obscures itself under the guise of science and 'greenwashing' (portraying a product or service as green or sustainable, when in fact it is not necessarily thus). There is still general confusion about the facts of climate change for many people and this is not always an accident. This is why it is essential that we practise 'safe scrolling' and information-gathering from credible sources, so as to ensure that we retain legitimate, scientifically based information in appropriate quantities from reliable, scientific sources.

Although this polarization is a threat to the effective communication of facts, trends are looking better for acceptance and action. A long-running programme of research in the United States has been charting public attitudes towards climate change for over a decade. It depicts the six different categories of Americans (ranging in engagement and concern, from most to least, as Alarmed, Concerned, Cautious, Disengaged, Doubtful, or Dismissive) and their views and participation in climate change. By 2020, the Alarmed segment had risen from 11% to 26% of the US adult population, while the Dismissive segment fell from 12% to 7%.[6] Quite simply, Americans are becoming more worried about global warming, more engaged with the issue and more supportive of climate solutions. People are beginning to listen to, and accept, the science of climate change. We know that when people accept this, they will be more likely to engage in pro-environmental behaviours.

The trend in the United States is being seen around the world. These research programmes are essential not only in informing our understanding of the landscape in terms of climate opinions but also in enabling us to have better and more effective ways of communicating about the climate across the cultural and political spectrum. And this is not just happening in Western democratic countries. The year 2021 has seen without doubt the largest global opinion poll on climate change. With over 1.2 million responses from 50 countries, the Peoples' Climate Vote[7] provides a truly incredible global timestamp. A mapping exercise of, for many countries involved, the uncharted waters of public climate opinion. What resonates is the broad consensus around the world of the urgency of addressing climate change, where recognition that we are facing a global climate emergency ranges from 58% (in Least Developed Countries) to 74% (in Small Island Developing States).[8] So all around the world, we know that at least 58% of people are on board with the urgency of the problem.

In the UK, work is being done to chart attitudes towards climate change across the cultural and political spectrum. The Britain Talks Climate[9] programme identifies seven distinct socio-political 'segments' (broadly speaking, from the most climate-engaged to the least, being Progressive Activists, Civic Pragmatists, Disengaged Battlers, Established Liberals, Loyal Nationals, Disengaged Traditionalists and Backbone Conservatives), with a view to better understand their views on the climate crisis. Encouragingly, though, they uncovered a wealth of common ground. In varying degrees, a majority across all segments agree that climate change is an issue that affects everybody (as opposed to merely affecting the left wing). Likewise, the segments agreed on the sources they most trust to deliver climate messages (in no particular order, climate scientists, environmental charities and, as a testament to the incredible impact of one single person on the awareness, understanding and sentiments of a nation, Sir David Attenborough). There are notable differences between groups, of course. For example, six of the seven segments subscribed to the idea that 'The effects of climate change are the same for everybody, regardless of race'. This is a problem because it does not consider the disproportionate climate impact on Black, Asian and minority ethnic (BAME) communities who are, for example, by virtue of their often living in more marginal or susceptible areas and with higher rates of unemployment, at a greater risk of the social and financial consequences of climate change – indeed, 'Minorities tend to live in places that are worst hit by the impacts of climate change – their poverty exacerbates their vulnerability'.[10]

'Climate change has an image problem'[11]

When we want to discuss or disseminate climate information, it's useful to be mindful that the language is appropriate for the audience, that we repeat these messages and that we utilize the power of stories, which bind us to the subject through our basic social instincts. We can also benefit from the power of imagery which, in the past, has communicated unhelpful messages, either because it featured 'distant' scenarios or graphic data, or because it highlighted merely the struggles of affected communities rather than what they have already achieved. There were few success stories told in pictures. For this reason, 'Imagery needs to embody people-centred narratives and positive solutions, and must resonate with the identity and values of the viewer – not just environmentalists. Only then can we truly drive engagement and promote positive action against climate change.'[12] This reminds us to use the *action* of 'liking and sharing' as useful and easy climate work that weaves into our lives and influences the narratives about the climate crisis and what we can do about it (see, for instance, Climate Outreach's 'Climate Visuals' library[13] as a fantastic online collection of powerful, positive climate photography from around the world). This also serves as an opportunity to pass the proverbial microphone, by showcasing helpful and empowering imagery and stories from MAPA communities, and other areas where amazing climate work is already underway. We can share these person-centred, solution-focused stories in our channels, harnessing the power of evidence-based imagery to help shift the image problem of climate communication.

VOICES

It's so important that the voices of people of colour are centred. That way the conversation becomes normalized for us too. If you search social media for climate anxiety, it's all narratives of white people – it's white voices – which are valid, of course, but it makes people who aren't as familiar with the term 'climate anxiety' think, 'Oh, this isn't something that I experience.' People most marginalized don't articulate, 'I have climate anxiety'. Those aren't their words, but they still feel the emotions. They don't have that in their vocabulary. That is why it is so important to speak about this, especially in spaces of people of colour because it helps to hear it from someone who looks like you, who is experiencing the same thing.

Raise awareness. Talk to people who are experiencing what is happening. It's easy to reach out to people, especially with social media now. If you are going to talk about the experiences of the people in MAPA, then it's important that you go out of your way to connect, or at least read something from their own voice. If you are in a space with someone from MAPA, amplify their voice. The best thing you can do if they can't do it for themselves is to silence your own voice and let them speak. So many people used to say, 'We are fighting for the voiceless – we are fighting for the people in MAPA.' We are not voiceless. You don't need to fight for us. Fight with us – be with us while we do this. We need you by our side. It's so important to know about what is happening in other parts of the world so that you don't just fight for what's good enough for Europe but what is good enough for the entire planet.

— *Mitzi Jonelle Tan, climate justice activist from the Philippines* —

Community and connections

It is clear that awareness about the climate crisis is increasing. From a psychological perspective, we notice that as people begin to engage with the gravity of the threat, they may rapidly go from a place of denial or disengagement to a state of profound understanding and acceptance. People often find this phase overwhelming. This can be intensely difficult for them as they may begin to close off, turning inward and being left feeling isolated. The Climate Psychology Alliance (CPA) highlights that this feeling of overwhelm increases if people feel they are alone in their fear, and this can lead to a sense of helplessness.[14]

This isolation can be further compounded for people when they perceive the apparent immunity others seem to have in comprehending the extent of the climate crisis. People tell us that they feel as if they are going crazy and struggle to make sense of the fact that no one else is as upset as they are (of course, there are millions of others who feel the same weight of the emergency, but it can be hard to remember this). Isolation can be hugely frustrating but, more dangerously, reinforce the narrative that this falls on us alone as individuals to fix.

When we take action and engage with others in a positive, impactful way, these feelings of helplessness, anger and isolation can begin to subside, along with anxiety and guilt, or at least allow us to connect with those who can relate as a shared experience. We can build connections and harness a greater level of mental sturdiness from our action-based support systems. In doing so, community and connection provide the mechanism through which we can translate individual action into wider systemic change.

Anne-Marie Bonneau, anti-plastic activist and author of *The Zero-Waste Chef*,[15] shared with us a little of her story, the emotions that have arisen from climate change and plastic pollution, and the solutions and mechanisms that have empowered her to make massive changes. As her community and voice have grown, she has created a ripple, or rather wave, effect of action to reduce or eliminate plastic waste.

VOICES

I've been concerned about it for a while but what really prompted me was the plastic pollution, reading about plastic pollution. I told my daughter I don't want to be a part of this and we started to reduce our plastic use. Community is so important. Find your people. I have met the nicest people. I have a group of friends who I've met through social media and my blog and we get together and sew produce bags that we make out of donated material. We then give them away at the farmers market and it reduces plastic that people are using at the market but it also gets people thinking. It all sends out ripples and lots of people have told me, 'Oh, I made this change and that change.' I've had so many people say, 'Oh, I don't like the plastic but I didn't know what to do' and then they think of other ways they can help. No one is going to stop using plastic straws and stop there. So I think cutting the plastic is like a gateway into further action.

— *Anne-Marie Bonneau* (The Zero-Waste Chef)

There is a lot to be said for meeting people where they are and finding common ground. Positive stories shared encourage us to remove our psychological defences. They bind us together. Stories bring the issue closer to home, while at the same time we feel motivated by the changes that we can make. We feel we have a system around us that provides support, empathy and a sense of collaboration.

Community serves us by letting us serve it. Helping others increases our sense of social connectedness, optimism and purpose. Helping others on their journey through climate emotions allows us to show off all the great skills we have learned and hype all the massive action we are taking. We don't have to be modest about our work. This behaviour is contagious. Generally, people would rather be shown what to do rather than told. A great way to influence others is by showing them how great this space is – how empowering this mentality feels and the strength we derive from this. Leaning into your own validation of your experience can help.

VOICES

Sharing validation of your experience with others really gives them the agency to take their own actions. It gives people the power to reclaim their own narrative and from that leads them to create their own view of how they see action and that leads them into becoming an activist. The activist as the educator, artist, photographer, model, videographer, that leads them to their journey of what they want to do with their activism, and in a way that they can do confidently for years.

Isaias Hernandez (@QueerBrownVegan) – environmental educator

We invite you to commit to contacting three people you can identify who would be helpful to speak with, if you are feeling the emotional weight of climate change. This might be someone you do climate work with, an employer, a friend or a family member. One of the wonderful things about our technological communication is that we are able to form connections all around the world – we don't even need to have necessarily met these people. It's wise to find someone who has themselves experienced these strong emotions. We encourage you to contact them now and ask them if it is okay for you to connect when future moments of difficulty occur.

By doing this in advance, you can let them know what you think might be helpful, whether that is meeting with them or engaging in some pro-environmental action with them, or it may just be that you need to be heard. By doing this, you also make room for others to feel more open to discussing their worries. You may end up getting a phone call or message yourself asking for advice. This is part of an emotional *safety plan*, so that when the waters get choppy, you know where to reach for that life vest. This isn't just about *feeling better*. It is about supporting yourself so that you can continue to support the planet. Networks facilitate systemic change.

Exercise: Strengthening my climate network

Who will I call?

1. . . . .

2. . . . .

3. . . . .

When you have called them or messaged or spoken with them, put a check mark next to their name to signal to yourself that you have people there who are available for you.

VOICES

I have a friend on every continent fighting for the same thing. How can we lose? It's almost mind-blowing – there are people in every country – that is inspiring.

— *Mitzi Jonelle Tan, climate justice activist from the Philippines*

Talking about climate change

As Climate Outreach reassure us:

> It's easy to believe that a 'successful' climate change conversation has to include dramatic moments of revelation, where someone who was initially sceptical about the problem changes their position. But this doesn't have to be the case. It could mean a two minute exchange on the bus with someone who has never talked to a stranger about climate change before; helping a friend who has been struggling with the issue shift closer to joining a local action group; or simply gaining the respect of someone who continues to disagree with you.[16]

Increasingly, greater emphasis is being placed on the importance of talking about climate change. For so long, the subject has felt too divisive, politicized and emotionally overpowering, and, as such, the safer thing to do has been to avoid the subject altogether, especially with friends and family. This has been further perpetuated by a collective lack of confidence in *how* to have these conversations. What tone should I strike? Should I appeal to their head or their heart? What do I say to someone whose attitudes and beliefs are different from mine, or even whose livelihood might be adversely affected (at least in the short term) by a move away from fossil fuels, for example? Work is being done to help empower people with the tools and ideas to begin having climate conversations, and in cases such as Climate Outreach's #TalkingClimate programme, even going as far as to develop networks, facilitate workshops and provide training to better promote helpful climate interactions.

We can also incorporate what we learned of the psychological defences that become activated when we receive climate information, and use solutions to these defences in our climate conversations. Per Espen Stoknes highlights some helpful approaches to overcoming some of the communication barriers of climate change:

- Make the issue feel near, human, personal and urgent.

- Use supportive framings that do not backfire by creating negative feelings.

- Reduce dissonance by providing opportunities for consistent and visible action.

- Avoid triggering the emotional need for denial through fear, guilt, self-protection.

- Reduce cultural and political polarization on the issue.[17]

This is again a moment to think about those naysayers that come to mind. When we think about what they are hearing and from whom, it becomes easier to understand that they are in fact victims, not only of their defences but also of motivated misinformation machines. When we think about this, it allows more space for empathy and the possibility to try to connect in other ways. To find common ground, find out what is important to people and get them to care and take action.

Narrative: the power of stories

─── **VOICES** ───

I recognized that one of my strengths drew from my experience of public speaking in high school. I realized that I have the ability to dilute climate terminology and spread it across my community. For me, it was easy to talk to friends about the climate crisis, instead of reading off a textbook definition that is generally presented within academia. I would often rephrase information and include my cultural and lived experiences and that helped a lot of people in my circles understand and engage with the climate crisis. The idea to reconnect personal narratives and personal experiences with others – it doesn't matter what race or class you are, people are interested in my work because I am able to bring that emotional side of me and that really activates other people to bring their emotional side. It allows them to say, 'Even though I have a different lived experience, I remember when I was a kid, we did this...maybe that has something to do with climate change.' That really resonates with a lot of people and that validation and connection over narratives allow them to build their own frameworks for action. The people I have met in my lifetime are doing such great work, there are so many environmentalists out there. The future is defined by collective thoughts rather than individual thoughts and so whenever I see myself thinking of a negative future of the climate crisis, I check in with my community. When we have these conversations, we share these thoughts and stories but we end up leaving with hope. If you push to have those conversations, we could have much.

─── *Isaias Hernandez (@QueerBrownVegan) – environmental educator* ───

The APA describes narrative psychology as a field that 'investigates the value of stories and storytelling in giving meaning to individuals' experiences – shaping their memory of past events, their understanding of the present, and their projections of future events – and in defining themselves and their lives'.[18] When we think about this in terms of global warming, it can be easy to detach ourselves from the messages, stories and connections that serve us. Rather, we may concede to a victim mentality of doom and gloom where all is lost. We can learn to challenge these narratives of hopelessness and instead cultivate identity, both on an individual and community level, that elicits an abundance of self-efficacy and harnesses the power of these stories to create positive change. Connecting through narratives is also a protective and solidifying coping strategy.

We see the importance of cultural and collective narratives as they bond communities' relationships with nature, history and resilience for adaptation in supportive ways. For example, when indigenous communities experience a disruption in their perceived connection to a place, and their practices within this place, they can feel a loss of cultural identity which can lead to low collective efficacy and be detrimental to their wellbeing. But because of the importance of storytelling within some of these groups, they harness strength and empowerment to adapt.[19]

We can identify the power of stories, but we need also to be aware of the social constructs that we might inadvertently place upon other groups. Because there is a narrative in media and social media depicting more devastating physical climate consequences for people in MAPA, often including indigenous groups, this can stigmatize indigenous people and cast them as being inherently vulnerable, in need of a 'white saviour'. We need to acknowledge that this (physical!) vulnerability to climate change is largely a result of the after-effects (and continued mentality) of colonialism. We must absolutely challenge the narrative that this vulnerability is often accepted as a given trait of indigenous communities, when in fact this comes from historic constructs and ecological conditions that are perpetuated, more so by Western, typically less by affected countries, and our disproportionate use of natural resources.[20] When met with climate events and a changing environment, this can have a compounding effect of trauma, essentially leaving these communities feeling that they have been robbed of their lands once again. This can result in negative mental health repercussions. This also reinforces that notion of helplessness in indigenous communities – that they lack resilience, perpetuating the myth that the former (or continued) oppressors need to interfere in order to help.

By listening, learning from, appreciating and centring the importance of these narratives, they may allow us to reconnect to the reciprocal relationship that we would benefit from having with our environment, rather than viewing natural resources as one-directional extraction deposits, as we typically do in capitalist economies. These individual and community narratives demonstrate the resilience of indigenous people, their connection to their natural surroundings and the power to draw attention to the rapidly changing environments in which they live. If we are to realize Albrecht's vision of the next epoch, the Symbiocene, we must listen to the powerful stories of natural oneness that come from these communities.

Thinking about our own personal narratives, let's ask ourselves who we want to be.

When we take ownership of our story, we can cast ourselves in the role of the good global citizen who listens to the science, takes responsibility as an individual and demands the necessary, large-scale, proactive response from governments. We can demand more equality. We can acknowledge the experiences of others. We can be the influencer. We can be the disrupter. We can be the connector of people. The listener. We don't all need to be Greta (as incredible as that would be!), but we can disrupt markets with what we ask of big corporations. We can protest through how we buy and vote. We can blossom from our personal histories, our lived experiences, use them to frame our now and our future in a way that emboldens us to take action. We can lean into what we are good at. We can cast ourselves as being committed to supporting the greater good, or the guardians of earth's species. As a leader who stands up for what we believe in. By defining the role, we buy into the character who is our ideal, effectively imagining this persona and then learning to cultivate and believe in the individual we want to become. We can then start showing up as that person. This helps to support the story that will transition us into the role we want to play, thus laying the foundation to modify our behaviours that resonate with this role and rejecting that victimhood mentality. It gives us a reason to stay committed and to believe. We cultivate a sense of hope and power.

No single character in a book does everything. Even the most lone of rangers has at least one companion. A mentor, a team. Each has its own distinct role to play. Revealing our personal narratives will allow us to consciously engage with how we are interpreting and navigating our situation, in the climate crisis and beyond. In order to investigate this, we need to ask ourselves some big questions. For this exercise, it is worth fencing off some time, as it may bring to the surface some revealing answers.

In creating your own narrative, it can be helpful to use what we are calling the STORY technique. Commit to writing this STORY for at least 15 minutes in order to deeply explore some of these questions. We suggest that you pick a few questions from each section, write them down and answer those that strike a chord for you. If you are used to journaling, this might feel natural. If this is new territory, gently guide yourself through these questions, honestly and with patience, as this could feel clunky or bring about a sense of vulnerability. That's okay; you are doing this for you.

Exercise: (re)Writing my STORY

S – Self

T – Trust

O – Ownership

R – Rewrites

Y – WhY

S – Self

How am I currently self-identifying? What stories and attributes am I cultivating? What is the backstory of my character? When do I present a hopelessness mentality? What does my unhelpful character look like, sound like? What changes could I make to my self-narrative to develop strength? What can my more helpful character look like?

Describe these as characters that are separate from yourself. Externalize them. Give your hero character a name and identity. Give them a voice, physical characteristics. Make them real. Developing the unhelpful character (the villain) might mean bringing awareness to, for example, negative self-talk or identifying when we are engaging in a hopelessness mentality. Developing this awareness will allow us to call ourselves out when we are self-sabotaging (setting ourselves up for failure – examples of this could be procrastination or overcommitting) and also when we are engaging in negative stories that don't support our wellbeing.

T – Trust

How do I cultivate a sense of self-trust, of self-belief? Do I think I am strong enough? When have I trusted myself in the past and been met with positive outcomes? Whom can I trust to support me in my narrative? How can I commit to trusting myself?

It may well be that, in creating this character, we need to describe them as we would *hope* them to be. Courageous in times we are not. A self-belief that we aspire to. Trusting ourselves to create the narrative that will spark joy and energy so that we can engage with our climate activism.

O – Outwards

Who else is in this story? How does my hero character relate to them? Are these others representative of who I want to be in the context of my wider community? Are there wider social narratives that fuel the actions of my character? Or impede them?

When we become conscious of the narratives that surround us and the collective narrative of our group, we can step outside of this for a moment and identify what is positive, but also, importantly, what is negative. What is perpetuating things like institutionalized racism or gender bias? This highlights how we are thinking about 'others' and if this is supporting or obstructing meaningful shifts in our societies towards more equality and climate action.

R – Rewrite

How will I review and rewrite my narrative when it needs to be adjusted? How can I do this from a place of self-compassion and pragmatism? How can I help to reframe the collective narratives around me? What are some of my strengths that I can lean into when I need to make adjustments to my story? How will I signal to myself when rewrites need to occur?

We are all figuring out what doesn't work. Reviewing and not being afraid to make rewrites within our narrative will keep us emotionally dynamic and nimble in our climate work. Forgive yourself slip-ups – they're part of your story and make you stronger.

Y – WhY

Why specifically does climate change matter to my character? Why is my character necessary in the outcome of this story? Why is it important that I share my story? Why is it important that I develop a sense that it is okay to shift towards positive optimism? Why can I let go of guilt?

It is important that we remember whY our story matters. There are so many potential stories to be told in a genre as wide-reaching as climate change. Be as specific as possible here. What does this character care most deeply about? Is this about conservation? If so, which species? Protecting habitats? Or equality and justice for marginalized groups, or the voiceless?

As we actively and intentionally construct our own narrative, we can build in resilience and self-compassion, along with determination and resourcefulness. We can spread this out and build communities that shift societal norms towards equality and ecological healing.

Chapter 7

Laying the Foundations for Action

In this chapter, we journey through the psychological prerequisites for climate action. The processes that don't necessarily appear in the books on 'what' to do to solve the climate crisis. Here we focus on the 'how'. If individual and systemic change were easy, automatic processes, this book wouldn't exist. Setting down strong emotional foundations can make the difference between a successful attempt at action and one that leaves us feeling dejected, with lower self-efficacy and a self-perpetuating sense of hopelessness and powerlessness.

Chapter 7 is about meeting yourself where you are, and proceeding from there.

Finding attunement

The very first step towards any lasting individual or collective shift, but notably in terms of climate change, is to achieve attunement. This holy grail of psychological states, explained best by psychologist Renée Lertzman,[1] is understood as being 'in tune' with all aspects of your psychological experience in that moment. Gazing inwards with compassion and curiosity. Not judging or resisting your experience. Noticing the presence of anxiety, frustration, motivation, whatever it may be. In our workshops, we devote time and attention to nurturing attunement for our participants, but in ourselves as well. One can't support attunement in others without first, as best we can, eliciting it in ourselves. Attunement is 'showing up' in its purest form. It can feel frustratingly evasive at times, though. Certain internal experiences – ironically, those that attunement is designed to shine a light on – can make it harder to access. Anxiety, anger, low mood all deplete our psychological bandwidth, making attunement harder to access. But not impossible.

Many of the following exercises are designed to encourage attunement. We journey through constructs such as scarcity, abundance, values, objects of care and mindfulness. All of these are designed to help ground in your current experience, without judgment, and anchor to what matters for you personally. As Paul Gilbert, the godfather of compassion-focused therapy reminds us, 'We did not choose to be born, nor the genes that made us, nor the kinds of emotions and desires that often operate within us... We just find ourselves here. We are all in the same boat.'[2]

These, we believe, are the necessary precursors to climate action, and so once you have had the opportunity to explore them, Chapter 8 is the point at which, according to Lertzman, your attuned self can begin to act.

Moving towards mindfulness

Mindfulness is a powerful tool that can help ground us in the present moment, as best we can, without judgment, but instead with openness and curiosity.

Rooted in the wisdom and practices of ancient Buddhist tradition, the early incarnations of mindfulness-based therapy centred around supporting people with chronic pain and depression. Both of these ailments are alike in the fact that psychological processes, such as negative thinking, worry and rumination, can not only distract from a person's attachment with the present moment but can create additional 'noise', acting as a volume dial increasing the intensity of the symptoms themselves.[3] We become distracted by the history of the problem ('This has affected so much of my life') and likewise its future ('Will this pain ever go away?'). If we imagine our thoughts, our attention, as being a pendulum that swings between the past, present and future, mindfulness allows the weight to rest, if only for a moment, dead centre. To stop oscillating and be still. To notice where we are.

More recently, as the practice of mindfulness has quite simply exploded across the Western world, so too has research investigating not only its effectiveness but how mindfulness might actually cause beneficial structural changes in the brain. It can preserve grey matter in older age,[4] thicken the hippocampus (an area associated with learning and memory) and even decrease cell volume in the amygdala (your FFF friend).[5]

Mindfulness allows us to plug in and be present in the moment. Not surprisingly, engaging with climate change can involve casting our minds back to past behaviours, or inactions, or instead ahead to a frightening future. But so much about our climate action will be about engaging with today, here, now. Here we offer the chance to dip the proverbial toe, with a brief mindfulness exercise to begin taking the emotional pulse, anchoring to the present moment, using the breath.[6]

Exercise: A brief 'breathing space' meditation

We recommend that you find a moment in your day to do this. A time when perhaps you are unlikely to be disturbed. To begin with, it helps to practise the breathing space in this way, by creating a specific time, so you have an opportunity to practise the technique. However, the breathing space differs from other, more formal, mindfulness practice in that it can be done in between activities. It is designed to be brief, a kind of 'mindfulness on the fly'.

The technique follows an 'hourglass' structure of awareness, in three steps: starting by capturing a wider sense of what exists in that moment, charting the emotional, mental and physical landscape, before 'zooming in' to the very narrow experience of the breath itself and, finally, expanding the awareness out again, from the breath, to include the body as a whole.

It can be helpful to audio-record yourself reading the instructions below, so you can be 'guided' in your practice. Likewise, a quick online search for 'breathing space meditation' will bring plenty of results, should you prefer a different voice.

Step 1: Bringing awareness to our thoughts, emotions and physical sensations

Begin by consciously positioning yourself in an engaged and open posture, sitting or standing up straight. Allow your shoulders to sink down slightly, and your facial expression to soften.

Allow your eyes to close if that feels comfortable. Ask yourself: 'What am I experiencing in this moment? What thoughts are coming to me?' Do not try to change them, just allow them to be. What emotions are here right now? Can you name them to yourself?

You don't have to feel as though you should judge them; just label what they are and notice that is your experience right now. It just is.

What physical experiences do you notice in your body? Are you feeling tightness or pressure or tension? Just allow yourself to notice these.

Step 2: 'Zooming in' to the breath

Now, shift the focus of your attention to the physical sensations of the breath.

Bring awareness to the rising and falling of the chest or abdomen, or of the sensation of the air as it enters and leaves the body through the nose or mouth. Do not try to breathe 'a certain way', or to change the breath. Just notice the breath exactly as it wants to be.

Each breath invites us to engage with this very moment. Here. Now.

If you notice your mind wandering, gently acknowledge this has happened, without having judged where it went and invite it to return to the breath.

Step 3: Expanding awareness out from the breath

On the next in-breath, zooming out now, allow the breath, and awareness of the breath, to radiate through the body.

Notice any areas where you might experience discomfort or tension and target your breath to channel there. Imagine you can breathe awareness into, and out from, every part of the body, breathing awareness right from the top of your head to the tips of your toes.

Then after a moment or two, expand the awareness out further, to include the room as a whole. The sounds around you, the feeling of the ground, the chair or the bed, beneath you. Notice how your body is interacting with the space around you.

Then finally allow your eyes to open, take in the visual information around you, but keep this sense of present-mindedness and connection as you go about the rest of your day.

Scarcity versus abundance

To experience scarcity means to be lacking in certain resources. This comes from feeling that we do not have enough to give, or that we lack the skills or time to respond accordingly to a situation. Emotional scarcity is what we can think of as a deficit in our psychological bandwidth. Sometimes in therapy and coaching, we refer to our 'stress bucket'. We all have a finite capacity with which to tolerate and respond to the demands of daily life. For much of the time, our capacity exceeds these demands, meaning we have a good 'buffer' remaining. When we are feeling strong and resilient, we have the sense that we could face great adversity, but if our stress buckets are full, and that buffer is gone, even a little extra drop can cause us to spill over and result in us entering the fight, flight or freeze mode. This is relatable to those 'final straw' moments we experience, and although they are often about losing our temper, they can also include breaking down in tears or switching off.

If we approach climate action from a place of scarcity, it may prove very difficult to maintain a sustainable response, because, emotionally, we feel we will never be, or do, enough and we perceive our actions as mere drops in a rising ocean. We did not produce global warming on our own, nor will we be able to fix the problem on our own. If we therefore approach climate change from a place of scarcity, we will quickly tire in our efforts and feel discouraged. We may even find ourselves so overcome by scarcity that this results in an inability to take any action at all, a sort of 'What's the point anyway?' response. Scarcity is not good for our mental wellbeing and certainly not good for the planet.

On the other hand, abundance signals having 'plenty of'. When we think of mental abundance, we think empowerment, growth, commitment, belief and positivity. This fortifies us and makes our climate action more sustainable. When we are able to notice our emotions and harness them to inspire abundance – thinking about the wonderful possibilities of all that we can do – we feel empowered and our self-efficacy can flourish. In doing so, we retain a sense of ease. We can become resilient, adaptive, dynamic – able to get back up when we get knocked down.

Mark Manson notes that 'Emotions are simply biological signals designed to nudge you in the direction of beneficial change'.[7] Our thoughts ignite our emotions. These emotions can give us the indication of whether we are feeling scarce or abundant. We attach emotional currency to our thoughts. For example, think of the word 'The'. Close your eyes and hold the word 'The' in your mind.

What do you feel? Nothing, right? It doesn't produce an emotion because there are no thoughts attached to it. It is simply 'The'. Now think of the word 'Mother'. Close your eyes and hold the word 'Mother' in your mind. What do you feel? This word can produce all kinds of emotions because we may associate the word 'Mother' with thoughts and memories. For some, 'Mother' may make us feel warm and joyous; for others, fearful or angry, or grief. This word comes with a thousand stories.

When we experience thoughts like 'Global warming is so awful. Why am I bothering? This isn't going to help! Nothing is going to work', we feel disempowered and insignificant. This infers scarcity. We can learn to tolerate the negative emotions without judgment. We can allow for them without trying to push them away or overcompensate for the scarcity we feel. Noticing our negative thoughts, that we are feeling scarce, is an indication that we need to amp up the self-compassion and self-care.

Likewise, thoughts that are rich in abundance like 'What an amazing project this is! We are all working so hard! I am going to get this done!' promote the positive emotions that more easily encourage us to *flow* in the direction of beneficial change. From this abundant space, it is easier to reinforce the behaviours and actions that feel good and support our climate work.

Creativity comes from abundance. We have the mental space to imagine what is possible. When we feel this abundance, we create solutions to support ourselves, our values and, ultimately, the planet. When we welcome obstacles with creativity, we see all the ways that we can help the planet and we feel robust in order to take those necessary steps. Abundant action is where we want to mentally anchor, in order to have the most productive impact. This goes hand in hand with acceptance of our emotional experience and a self-compassion that allows us to continue onwards in an emotionally sustainable way.

It's really hard to scream 'Be more abundant!' at ourselves when we're feeling scarce. This will likely do more harm than good, by reminding us of the 'abundance deficit' we are experiencing in that moment. Instead, it may be about learning to identify what scarce *feels* like for you and experimenting with those moments. Be curious. Is there some activity you can do, or person you can contact, that might gently nudge you to feel more abundant? If not, might it instead be about knowing the need to down tools for a moment? Be patient, care for yourself, replenish the buffer zone in your stress bucket, and delay your actions until you feel abundant again.

The role of self-efficacy

Another determinant of climate action is self-efficacy. We have used this term throughout the book thus far, because it interrelates with so many aspects of our psychological response to climate change.

But what exactly is self-efficacy and why might it be important? Eminent psychologist Albert Bandura first described self-efficacy as 'people's beliefs about their capabilities to produce designated levels of performance that exercise influence over events that affect their lives'.[8] So, essentially, if we believe that we can achieve the goals we set ourselves, and that by achieving these goals we can affect a wider situation, we are much more likely to set about doing so. We can see quite quickly how this can create a positive feedback loop (Figure 7.1). If we are more likely to set goals, we are in turn more likely to achieve tasks and see a gain from doing so. This positive reinforcement means we are more likely to continue setting goals in that area of our lives.

Figure 7.1 High self-efficacy reinforcement loop

The same can be said for lower self-efficacy (Figure 7.2). If we feel unmotivated or disempowered, we don't set goals; therefore, a process of negative reinforcement can occur whereby we don't have the opportunity to experience any tangible changes or benefits in that area of our lives. This creates an equal and opposite cycle and reinforces the belief that we have no control or impact over events.

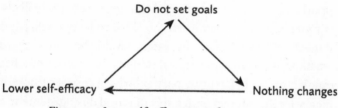

Figure 7.2 Low self-efficacy reinforcement loop

There are two fortunate things to hold in mind here. First, self-efficacy can be improved upon in well-evidenced ways: by setting meaningful and achievable goals (more on this later in the book), and also by holding the bigger picture in mind (e.g. avoiding becoming overly myopic or becoming disheartened by specific goals that weren't achieved). People acting with high levels of self-efficacy will continue with a project despite setbacks, with a conviction that it will come together regardless. They will maintain a positive self-narrative, too, by perhaps reminding themselves of their past successes, goals achieved or times when setbacks or accidents actually produced a positive result. Failure is reframed.

Self-efficacy is known to vary from one area of our lives to another. If you are feeling less motivated or action-focused in relation to climate change (suggesting low self-efficacy), it is likely there is another area of your life (perhaps work, friendships, hobbies), where you experience a greater level of self-efficacy. You can then use this as a template, which can be applied to your response to climate action.

In relation to climate change, we can think of efficacy at different levels: self-efficacy (belief in the possibility of climate action) and response efficacy (belief that action will be effective). But efficacy also exists in relation to wider systems, whereby a person's sense of collective and governmental efficacy will determine their engagement with climate action, and active support for pro-environmental policy.[9] Put simply, if we believe that individual action is possible and likely to be effective, we're more likely to do something; also, if we believe that our communities and elected politicians have the will and opportunity to effect meaningful change, we are more likely to engage more actively at that level, too. This collective efficacy is a crucial mechanism, therefore, in system-level behavioural change. Bandura himself has for a long time described the role of self-efficacy in people's climate behaviour, particularly in terms of how low self-efficacy might interact with a sort of moral disengagement, becoming detached from one's sense of what is right, to allow existing non-environmental choices and actions to continue.[10]

Evidence suggests that self-efficacy is a good predictor of climate change adaptation and pro-environmental behavioural intent. However, it is not only influenced by intrapsychic factors, but also, particularly in areas where people's livelihood is directly threatened by climate change, systemic factors such as technology, material resource and infrastructure also predict their sense of efficacy.[11]

Objects of care

Objects of care relate to the things that we hold in our minds as being the most important, most emotionally resonant and over which we feel the greatest sense of guardianship. They are deliberately and purposefully specific and are likely shaped by our prior experiences – where we live, grew up or perhaps hold the fondest memories of travel or immersion into nature. In relation to climate, these objects could be people (our children, other citizens of the world) or perhaps emblems of the natural world (forests, certain animal species, wetlands). That the World Wide Fund for Nature (WWF) chose the giant panda as its charity logo was no accident.

By locating conservationism very specifically and tangibly on such a recognized and well-loved species, it soon became a symbol of guardianship around the world. Objects command our attention and vigilance, and we are motivated to protect them over all other things. We attune to objects. It has been suggested that in relation to climate change, there is a particularly strong emotional response that incites action when one perceives their objects of care to be under threat, but this level of care generates greater environmental thinking, and even predicts levels of support for green policy.[12]

During one of our workshops, we asked participants to tell us why saving the planet was important to them (to identify their objects of care). Many said for the generations to come; that they felt it was their duty to ensure that their children and children's children lived on a viable planet. One participant raised their hand and said that they felt compelled to engage in climate action to honour their grandparents. This person said, 'They did not survive two world wars to see us destroy this beautiful planet.' This highlights that people are driven by deep wishes to live a life that is congruent with the way they believe the world should be. For this person, it was about honouring elders. It could indeed be that we locate our objects of care as people. It could also be our spiritual connection with nature, it could be that we value social justice and see the implications of climate change and poverty, but it's important that we identify what is it that makes climate change matter for us. We have different things that motivate us to act and feel empowered. When we tap into what, and who, these objects are, it is much easier to engage in activity from a position of conviction and determination. We embrace our emotions and look outward as they can fortify us. We do the work because it matters to us.

So how do we locate our objects of care? In a moment, we want you to immerse yourself in an imagined future scenario, which may help to illuminate a little of what matters to you. We do this by waving the 'magic wand' or, in this case, the 'ecological magic wand'. Doing this in a quiet space when you are feeling resilient and focused, visualize this ideal future and take some time to respond to the questions in your mind. We encourage you to perhaps write these answers down. Getting thoughts and ideas out of our heads and on to paper creates mental space and can allow us to zoom in on themes and ideas that may not get the attention they deserve in a cluttered stream of thought.

Exercise: Imagining an ideal future

Take a moment to imagine that individuals, big business and governments began today to make all the necessary changes to achieve net-zero emissions (you may notice yourself resisting this, but try to just imagine this success). You now find yourself in the year 2040. People woke up and saw what needed to be done. The air is clearing, the oceans are repairing, there are huge rewilded areas that have allowed once-endangered species to thrive. The soil is regenerating and there is a fundamental respect for our planet. Humanity is more equal and more thoughtful. Imagining yourself in this world...

Where are you?

What does it feel like to be here?

What are you doing?

Who are you with?

Who or what has been saved?

What matters to you in this place?

What made all of this climate action worth it?

Anchor on to the feelings and connections in this place. Use this visualization as a kind of lighthouse when you feel you have lost your way. Reconnect with this place, the objects that inhabit this place, this goal, this ideal future, to remind yourself of some of the reasons you are committed to climate action.

Values

Our values reflect what we find important in life. Values are individual and can change throughout a lifetime. They show us how we want to engage with the world, with the people around us and with ourselves. Values navigate us through our daily lives. You might think of values as being a constant compass, silently guiding your thoughts, emotions and behaviours.

Have a think about *what* you value. If this feels tricky, a good way of teasing this out is to ask yourself, 'How would I *hope* that a good friend describes my positive attributes?' Still finding it hard? At the risk of being morbid, try asking, 'If someone who loved me was giving my eulogy, what would I *hope* they would say?'

Values and strengths are closely associated. If our values represent our ideal, or best, self, our Signature Strengths are those that instead represent the best of our actual selves, all things considered. It can feel awkward to think in this way. Social law dictates we must show humility at all times. Also, negative self-talk can attempt to quash any efforts to locate one's own strengths: 'Oh, you aren't really as good at that as you think.' But, as best you can, approach the following exercise openly and honestly. It's okay to have strengths! Everybody does. Take a moment to identify these in you; later in the book, we return to how you can translate your values and Signature Strengths into meaningful climate action.

Exercise: The Signature Strengths Survey (produced by the VIA Character Institute)[13]

In this exercise, read through the 24 Signature Strengths and tick any that you feel apply to you. Which ones best describe the essence of your character? What do you bring to the table? The authors of the survey suggest that you describe the person you are, NOT the person you wish you could be. Also, think about your life in general, not how you behaved in one or two situations.

Essential Strength?	Character Strengths
	1. Creativity: You are viewed as a creative person; you see, do, and/or create things that are of use; you think of unique ways to solve problems and be productive.
	2. Curiosity: You are an explorer; you seek novelty; you are interested in new activities, ideas, and people; you are open to new experiences.
	3. Judgment/Critical Thinking: You are analytical; you examine things from all sides; you do not jump to conclusions, but instead attempt to weigh all the evidence when making decisions.
	4. Love of Learning: You often find ways to deepen your knowledge and experiences; you regularly look for new opportunities to learn; you are passionate about building knowledge.
	5. Perspective/Wisdom: You take the 'big picture' view of things; others turn to you for wise advice; you help others make sense of the world; you learn from your mistakes.
	6. Bravery/Courage: You face your fears and overcome challenges and adversity; you stand up for what is right; you do not shrink in the face of pain or inner tension or turmoil.
	7. Perseverance: You keep going and going when you have a goal in mind; you attempt to overcome all obstacles; you finish what you start.
	8. Honesty: You are a person of high integrity and authenticity; you tell the truth, even when it hurts; you present yourself to others in a sincere way; you take responsibility for your actions.
	9. Zest: You are enthusiastic toward life; you are highly energetic and activated; you use your energy to the fullest degree.

	10. Love: You are warm and genuine to others; you not only share but are open to receiving love from others; you value growing close and intimate with others.
	11. Kindness: You do good things for people; you help and care for others; you are generous and giving; you are compassionate.
	12. Social Intelligence: You pay close attention to social nuances and the emotions of others; you have good insight into what makes people 'tick'; you seem to know what to say and do in any social situation.
	13. Teamwork: You are a collaborative and participative member in groups and teams; you are loyal to your group; you feel a strong sense of duty to your group; you always do your share.
	14. Fairness: You believe strongly in an equal and just opportunity for all; you don't let personal feelings bias your decisions about others; you treat people the way you want to be treated.
	15. Leadership: You positively influence those you lead; you prefer to lead than to follow; you are very good at organizing and taking charge for the collective benefit of the group.
	16. Forgiveness/Mercy: You readily let go of hurt after you are wronged; you give people a second chance; you are not vengeful or resentful; you accept people's shortcomings.
	17. Humility/Modesty: You let your accomplishments speak for themselves; you see your own goodness but prefer to focus the attention on others; you do not see yourself as more special than others; you admit your imperfections.
	18. Prudence: You are wisely cautious; you are planful and conscientious; you are careful to not take undue risks or do things you might later regret.

cont.

Essential Strength?	Character Strengths
	19. Self-Regulation: You are a very disciplined person; you manage your vices and bad habits; you stay calm and cool under pressure; you manage your impulses and emotions.
	20. Appreciation of Beauty & Excellence: You notice the beauty and excellence around you; you are often awe-struck by beauty, greatness, and/or the moral goodness you witness; you are often filled with wonder.
	21. Gratitude: You regularly experience and express thankfulness; you don't take the good things that happen in your life for granted; you tend to feel blessed in many circumstances.
	22. Hope: You are optimistic, expecting the best to happen; you believe in and work toward a positive future; you can think of many pathways to reach your goals.
	23. Humour: You are playful; you love to make people smile and laugh; your sense of humour helps you connect closely to others; you brighten gloomy situations with fun and/or jokes.
	24. Spirituality/Sense of Meaning: You hold a set of beliefs, whether religious or not, about how your life is part of something bigger and more meaningful; those beliefs shape your behaviour and provide a sense of comfort, understanding, and purpose.
	None of these characteristics is more essential to who I am than any of the others. (Remember, you should choose this option if the strengths are all equally essential to you, NOT because you think they should be equally essential.)

Final Step: Review the strengths you checked. Do any of these strengths stand out as more important to who you are than the others? If so, put a second check in the box next to those strengths.

Challenging our (climate) thinking

Stepping stone thoughts

There is nothing either good or bad, but thinking makes it so.

(Shakespeare's *Hamlet*, Act 2 Scene 2)

We use a stepping stone model that can help us to modify our thoughts so that they are useful to us and result in helpful actions for the planet.

We are not asking you to go from thinking 'the earth is dying' to 'I'm sure it'll work itself out' – by no means at all! Nor are we wanting you to go from feeling anxious to carefree. We want to harness these emotions and enable them to drive us towards our objectives. What we hope is that we can model our thinking to use both negative and positive emotions to channel towards productive, sustainable solutions. So how can we do this? Think of our thoughts as stepping stones. If we imagine that we are currently standing on a stone (a thought), this will give us cues about our 'footing'. If we are standing on a thought that feels negative, our grounding can feel wobbly. Staying with this metaphor, we see a stone further up the path that we want to arrive at. It wouldn't be possible to take one big leap to this stone; instead, we need to hop on a few others along the way to reach it. By moving one stone at a time, we can begin to modify our thoughts and the feelings that they produce in order to arrive at a place that feels less wobbly.

So, for example:

Fact: There is a climate crisis.

First stone: 'There is too much that needs to be done to stop the climate crisis.'

Feelings: Overwhelm, anxiety.

We hop to the second stone: 'There is so much that needs to be done to stop the climate crisis.'

Feelings: Scarcity, worry.

Third stone: 'There are a lot of things that can be done to stop the climate crisis.'

Feelings: Opportunity, growing optimism.

Fourth stone: 'There is so much that I can do to help stop the climate crisis.'

Feelings: Motivation, commitment, abundance.

Here, the fact remains the same ('there is a climate crisis'), but our interpretation of this fact results in how we feel and what we do about it.

Exercise: Stepping stone thoughts

Think about a recent time that you have had a negative reaction to news or information you have received about climate change. Just for a moment, notice that feeling in your body. Where is that physical sensation? The pit of your stomach? Tightness in your chest or throat? Just take a moment to feel that and focus on breathing into that space in your body. Now coming back to the image or fact that provoked that feeling, identify what thoughts came to you – your first stone. You may need more or fewer stones, but what is important is that you arrive on a thought that feels authentic, but also promotes the best outcomes, emotionally and for climate pro-activity. Remember, we need positive action and positive action comes from positive thinking.

Fact (or triggering information):

. . . .

First stone:

. . . .

Feelings:

. . . .

Second stone:

. . . .

Feelings:

. . . .

Third stone:

. . . .

Feelings:

. . . .

Fourth stone:

. . . .

Feelings:

. . . .

Overcoming thinking traps

As you went through the stepping stones, you may have noticed certain thinking traps being evident, particularly in the initial stone or two. Seldom do these traps really serve us, in the way that we have suggested our defences and emotions might.

In many respects, the first (and arguably the most powerful) tool in overcoming thinking traps relating to climate change is to simply observe their presence. Better the devil you know.

Then apply a degree of distance between yourself and the thought – for example, by labelling the thought. Instead of 'There is nothing I can do to make a difference to climate change', try 'I notice I am having the thought "There is nothing I can do to make a difference to climate change", which might be a clue that I have fallen into the "all or nothing" thinking trap', or perhaps 'Am I feeling powerless, and therefore the emotional reasoning thinking trap is convincing me that there is nothing I can do?' Here we revisit the list of thinking traps from Chapter 5. As best you can, remain curious and open to your thoughts, and begin asking yourself the questions below, once you have identified that a trap has become activated. Use these questions to help shift perspective and hold the thinking traps to account:

The mental filter

For example: 'Nothing I am doing is making a difference'
Am I only allowing certain evidence to make it through? Noticing only my failures, or those of the people around me? Is there a counterargument to this thought and what evidence is there for that?

'All or nothing' thinking

For example: 'If I don't work tirelessly and
constantly to fight climate change, I've failed'
Am I using words like 'always', 'never', 'everything' or 'nothing'? Is it accurate to do so? Are there any exceptions to this? What do the shades of grey look like? Is it true that *sometimes* this is the case, but not always? Am I extrapolating? Seeing a pattern as absolute based on a single event?

Emotional reasoning

For example: 'I feel powerless in the fight against
climate change; therefore there is nothing I can do'
Am I confusing a *feeling* with a *fact*? Or basing evidence of truth
on an emotion?

Personalization

For example: 'This is all my fault'
What would I tell a friend who was experiencing this thought?
Am I putting an unfair amount of responsibility on to my own
shoulders, or those of somebody else?

Magnification and minimization

For example: 'My successes are so trivial'
Is there good evidence for the prediction I am making? Am I either
catastrophizing or downplaying? Have I made similar forecasts in
the past and were they accurate?

Along with this 'cognitive' perspective-shifting approach, you
can also look to certain behavioural techniques to help shift the
thinking traps. You'll notice that much of the shifting comes from
evaluating the evidence of the thought. This works particularly well
with the thinking traps because they represent certain distortions
in our thinking. So when you are evaluating the evidence for a
certain thought (e.g. 'There is nothing I can do to help climate
change'), write this thought down, and then record all the evidence
for and against. Be as specific as possible. Pay attention to examples
throughout the day of decisions you are making that are beneficial
to the planet and record them somewhere. Create a compelling
counterargument to the unhelpful, and unevidenced, thought.

Of course, particularly in the case of climate change, your
thoughts, concerns and predictions may be entirely accurate, or you
may indeed be responding to factual information. This is particularly

true when receiving climate news from trusted sources. However, there are many ways that we interact with climate change that do become vulnerable to these thinking traps, most notably in terms of what we, and others, can do to address the problem.

STOPP technique

The STOPP technique is about practising the art of harnessing that space, that split second, between stimulus and response. By noticing one of life's triggers, and just hitting pause, if only momentarily, you allow yourself to respond, rather than react. The STOPP technique is a way of supporting yourself to move from System 1 to System 2 thinking – from an automatic to a slower, more deliberate and considered response. In doing so, it is said, you can regain a sense of self-efficacy. 'I can choose how I respond. That will always be within my control. In this, I have my power.'

The technique is used to good effect by people who want to regain control over problematic anxious reactions, as well as anger and persistent negative thinking. The process can take a minute or two, or it can take merely a few seconds, particularly after having practised several times. It is true, though, that we often notice we have been triggered *after* we have reacted in an unhelpful way. 'STOPPing' takes patience and perseverance and you are bound to forget, or miss an opportunity, sometimes. Allow yourself this.

Once you notice that you have been triggered:

S – Stop!

This might sound obvious, but it makes all the difference. Often, people actually find it helpful to include some physical movement – a click of the fingers or tapping the counter with the palm of their hand –some short, sharp signifier of the need to stop. Or even just saying 'stop' either aloud or silently to themselves, to signify the need to do so. To pause and take stock.

T – Take a breath

Breathing is something you do 20,000 times a day, but make this one count! Use this next breath as an opportunity to ground yourself. Revisit the skills you practised in the breathing space exercise. Use the breath as an anchor. Breathe slowly and purposefully in through your nose. Noticing the sensation of the air as it moves through the body. Then as you exhale, slowly through the mouth, again notice the air as it leaves. Notice the successive rise and fall of the chest or abdomen. Notice these clues the breath leaves. These signs of its presence.

O – Observe

Continue to hold your attention on the breath, but then expand outwards to include the wholeness of your experience in that moment. Observe what thoughts are there. How are you feeling in your body? Do you notice any urges to react in a certain way? Any resistance? At this stage, do not try to change anything or judge your experience at all. Just take in the mental and emotional landscape, non-judgmentally. Remember, we all just find ourselves here. This is the way things are in this precise moment.

P – Perspective-take

At this point, you can begin to explore the role of perspective, gently asking yourself questions that allow for a fresh standpoint: 'Have I thought this in the past?', 'Is there good evidence for this thought?', 'Is this suggestive of a thinking trap?', 'Is this thought serving me (and the planet) right now?', 'What would I tell a friend in this scenario?' Remember, this is not about judging yourself for your experience, but instead holding your thoughts and feelings a little at arm's length for a moment. Viewing them from the outside and deciding how best to relate to them.

P – Practise what works, and proceed

Finally, dig into your toolkit. How are you going to respond? What have you done in the past that has helped in this situation? This might, and indeed often does, involve *delaying* your response for a while. Give yourself permission to come back to the problem when the intensity of the trigger has subsided. What you decide upon may be problem-focused (i.e. something that directly targets the triggering issue or climate information) or emotion-focused (something to help instead with the impact the trigger has had on you). Either way, by moving through these steps, it gives you a real chance of responding in a way that best meets your needs and keeps you sustained.

Why self-care matters

— **VOICES** —

I...began to really internalize that I can't help the world if I am not taking care of myself and began to view self-care as radical, important work rather than something to feel guilty about.

— *Former client (K)* —

There is a reason that inflight safety information instructs you to put on your own oxygen mask first. How can you help others if you can't breathe?

It's fundamental to establish the importance of self-care. This way, we can manage our climate emotions, as much as possible, from a place of physical and mental wellbeing. Most of us have experienced that day or two before a cold or flu where we just feel awful. We are tired, grumpy, we don't sleep well and we aren't sure yet why. We might manage this by cancelling plans with friends and opting for a quiet night in, or we might push ourselves out and feel worse for it. Our body is telling us that something isn't right and we are unconsciously experiencing and acting upon this signal.

This isn't too dissimilar to what can happen when we are feeling psychological discomfort. We may drop the ball on ourselves. We may feel distracted or demotivated, believing that our own physical and emotional wellbeing is inconsequential or unworthy, in the grand scheme of all things climate. Or we just simply forget to do the things we know are helpful. But when we are feeling psychologically wobbly, what is imperative is that we heighten our devotion to self-care.

'But how can I think about myself when the ice caps are melting?'

Let's take it out of the climate context for a moment and ask ourselves, 'What would a star athlete do?' It is the Olympic Games. Tomorrow morning is *the race* – the one that will make or break a career. The night before, does a winning athlete say, 'Everything is on the line for me tomorrow. I think I'll hit McDonald's and Netflix until 3am'? No, they don't. They understand and honour how essential it is to give themselves their best shot, to bring their full game, and that starts with being physically and mentally on form. If an athlete is willing to do that for a sport, we think your efforts in saving the planet deserve at least that same level of self-care.

What does self-care look like in relation to climate change? It might be helpful to first look instead at what it doesn't look like. When we try to resist emotional discomfort, it is like holding a beach ball underwater. The more we push it down, the more it will pop up with greater force. Likewise, the process is effortful and draining, and we become slowly exhausted over time. Negative strategies or habits function as distractions so that we can resist engaging with our pain. Holding it underwater as best we can. This could take the form of social media overuse or information overloading, unhealthy lifestyle (overeating, drink, drugs), avoiding, procrastinating, or, conversely, overcommitting or working too hard. These often bring a short-term sense of relief, which is why they can be so appealing. However, what unites these tools is that they end up recycling the stress in the longer term. How are we going to do our part to save the world if we aren't caring for our needs?

Exercise: My less helpful coping strategies

Identify some of your own unhelpful coping tools that can, at times, negatively impact on your physical and mental wellbeing:

1. . . . .

2. . . . .

3. . . . .

4. . . . .

5. . . . .

By identifying some of the unhelpful ways that we cope, we can learn to take better care of ourselves by challenging or removing the urges that superficially dull our pain or discomfort. What measures could you take to start shifting these? For example, we may be overcommitting, whereby we are constantly engaging in climate work and never giving ourselves a moment of reprieve. Trust us: this doesn't work in the long term. Or perhaps you notice evenings when you are scrolling through your phone, seeing climate content and, before you know it, you realize that two-thirds of that bottle of Pinot has mysteriously vanished. We don't need to deprive ourselves of things that we enjoy in moderation, but when it becomes detrimental to our wellbeing, we need to hold ourselves accountable and nurture more healthy strategies of self-care.

We can benefit by reflecting on other areas in our lives in which we are putting ourselves on the back burner, whether it be in our relationships, our health or our climate work. When we notice this, we can begin to recognize how important it is for us to permit ourselves to create balance, which will allow for more emotional stamina in all areas of our lives.

A good way to start is by gauging how balanced your life is feeling at the moment. If, on a scale of 1 to 10, we were to rate the quality of the aspects of our lives below, we can see where we are feeling satisfied and fulfilled and where we could do with spending a little more time to develop.

Exercise: Gauging your whole-life balance

Looking at the past month, how would you rate the quality of the life areas below? (1= least optimal, 10 = most optimal)

Personal relationships: Do you feel you are spending quality time with the people that are important to you?

Career and finance: Are you enjoying your work and developing the way you would like to? Are you achieving your goals? Are you feeling good about your finances?

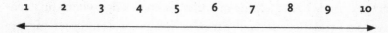

Health: Are you taking care of yourself? Exercising, eating and sleeping well?

Self-development: Are you feeling connected spiritually or ethically to your values? Are you engaging in mentally stimulating activities or learning?

Climate contribution: Are you feeling impactful and productive in the work you are engaging with to help the planet?

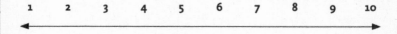

1   2   3   4   5   6   7   8   9   10

Once you have completed this, have a look at where the numbers are higher. What is it that is allowing you to achieve so much in that life area? Remember that you should not be aiming for 10 across the board. That would be exhausting. We want to ensure that we are being reasonable about what we expect from ourselves. We want to establish the notion of 'enough'.

There will be times when these numbers fluctuate. None of your responses above will be static. They will shift from one day to the next, depending on the demands life is placing on you in these areas. So, accordingly, ensure that you are flexible in terms of what your level of contribution needs to be in each area. Quite simply, you will have more to give at one time than another. 'Sustainable action' needn't have a finite quantity. Acknowledge when you need to do a little more for self-care and when you can direct energy outwards. In particular, when you think about how fulfilled you are feeling in terms of your climate work, it is important to remember that you will not be able to save the planet single-handedly. Thinking about what 'enough' means for you in order to keep a balance but also maintain motivated action will help you in managing your time well.

You can also pay self-care forwards, perhaps at work or with friends. We have heard wonderful stories about WhatsApp groups over the past year – disparate, COVID-torn friendships coming together to support one another. Even messages like 'Hope you're all okay' or 'I'm here if anyone wants to chat' can resonate in ways we aren't always aware of or don't always appreciate. Perhaps, we can apply the same to climate (youth climate activist groups are particularly fantastic at supporting each other already).

We can promote other people's self-care by:

- creating a space to talk

- recognizing and accepting their own climate emotions

- locating their climate values and guiding their actions by them

- helping them to focus on specific projects together and supporting them to 'say no'

- connecting together, and with nature

- noticing any unhelpful coping strategies (doomscrolling, avoidance, rumination) and offering support

- helping them to develop their own self-care plan

- sharing and championing their successes.

Getting outside (it's only natural)

It's time to overcome our nature-deficit disorder.[14]

Richard Louv's term (not formally recognized as a mental health problem at all, but rather a collective syndrome representing a detachment from nature and the ill-effects this can bring) captures a growing recognition of, and concern for, the pull of our modern lives away from the natural world.

Governments around the world are increasingly structuring legislation to promote interaction with the natural environment, including minimum targets for green space areas in cities and urban development. Likewise, various forms of 'green prescription' are being offered by health professionals as a way of promoting both physical and mental wellbeing through immersion with nature.[15] In the UK, for example, there is currently a widely supported application being considered for a new high school subject encouraging young people outdoors and reconnecting with their ecological surroundings: the GCSE in Natural History.[16]

Around the world, humans have, in so many profound ways, described our oneness with nature, with its own global taxonomy – words to best describe feelings and experiences. In German, we may engage in *Waldeinsamkeit* (wandering alone in the woods), or try *Shinrin-Yoku* (Forest-bathing) in Japan. *Dadirri* describes the Aboriginal Australian practice of 'deep listening', often in concert with the natural world. *Aloha 'Āina* describes the Native Hawaiian 'love of the land', a sense of being connected with all living things.

We can connect both directly and vicariously with nature. That is to say, we can experience an immersion into the natural world and also share the complete, immeasurable and precious joy of, say, writers or poets. To open the pages of Durrell's *Corfu Trilogy*[17] or McAnulty's *Diary of a Young Naturalist*[18] is to invite a heady, multisensory experience. A well-needed reunion with nature. We needn't always, however, travel far. McAnulty's writing particularly reminds us that we can meet with the natural world anywhere. In the trees surrounding our homes, or in the cracks in the pavement.

While, of course, being in nature is preferable, in times when direct immersion isn't possible, we can also visualize time spent in the natural world to surprisingly good effect. Here we invite you into another visualization exercise, not instead of heading into the natural world, but rather 'as well as'.

Exercise: 'Return to nature' visualization

This is a useful meditation technique to reconnect with nature and your environment, even if you are sitting in your living room. When we talked about *attunement*, we invited you to engage in a breathing space exercise. This visualization draws on that skill and includes an imaginary immersion into nature. When practising this meditation, try as best you can to imagine the fullness of the sensory information, the sights, sounds, smells and tastes you experience, and the physical interaction you have with this place.

You can choose this place in your mind. It is yours. It may be somewhere you have visited in the past – a forest, beach or desert. It may be somewhere that you completely invent in your mind's eye. Or it could be a hybrid place, inspired by somewhere you have been or seen images of, and augmented in your imagination to become a perfect place. Don't worry if the place takes a while, even several attempts, to become whole in your mind. You can develop and refine it over time. You may even have more than one place, and you choose where you would like to visit each time you perform the exercise. Spend a moment now thinking about what this place *could* look like for you.

This is a meditation that you can come to whenever you feel it would help you create a space for some calm, but not only when you are feeling anxious. This should serve more as a script. As we have previously suggested, it may help to make an audio recording of yourself, or somebody else, reading the script below, speaking slowly and allowing for reflective pauses. If this isn't possible, slowly read through this when you are feeling calm, ensuring that you allow time for breathing, so that you eventually become comfortable taking yourself through it.

Coming to this mediation, allow yourself to find a comfortable, dignified position, lying down or sitting in a chair. Adjust anything that feels uncomfortable and slowly bring your attention to the breath. Gently close your eyes. Take a moment to breathe. You aren't trying to change your breath, just noticing your breath as it enters...and leaves...the body.

You might have come to this practice because you are experiencing difficult thoughts or perhaps to cultivate a sense of peace or connection. Whatever your reason for being here, take a moment to thank yourself for supporting your wellbeing.

Breathe in, breathe out.

Again, notice your breath and, this time, consciously slow your breath to take a long, deep exhale, pushing the air out of your body as much as you can, and then slowly, through the nose, take a long, slow inhale. Breathe and pay attention to the breath.

Now you are going to gently guide yourself toward your safe and peaceful place. Take a few imaginary steps forwards into this place. Imagine yourself in this place of complete natural beauty.

Taking another few breaths in, you are going to focus on your senses in this place.

What does this place look like for you? Are you in the mountains? Is there snow? A beach? By water? Or trees? Where is this peaceful place? Look at the flora and the fauna. What is your favourite thing to look at here?

We invite you to sit or lie down in this place. Now notice what you can feel against your skin in this place. What does the ground beneath you feel like? What is the temperature like? Is there anything cooling? Water or cool sand beneath you? Or, likewise, anything warming your skin? Where you are grounded, what does it feel like? Can you reach out and touch something? Feel sand through your fingers or water in a stream. Imagine you can really hold the texture for a moment.

Breathe in, breathe out.

Next, shift the focus of your awareness to your imaginary sense of hearing. What sounds can you hear? Is there a breeze? Birds? The sound of water? What noises let you know that you are in this safe, beautiful place? Or perhaps it is the relative absence of sound that you find so relaxing. The silence of the natural world.

And if your mind starts to wander or takes you away from this space, notice this, without judgment, and gently return your concentration back to what it feels like to breathe in this place of beauty. You are held safe here. You are protected and nurtured by nature.

Breathe gently while listening to these natural sounds.

Now, on the next inhale, shift to what smells you find here.

As you sit in the quiet space, filled with the wonders of nature, what scents are there? Perhaps these are smells that take you back in your memory, to other times you have been here. You can imagine smelling a flower, or the saltwater from the sea, or the richness of the woods. These scents calm you and connect you emotionally to this place.

You then notice your favourite thing to eat in this place. You pick it up and take a bite. What flavours are you enjoying in your safe, calm space? Try to really imagine tasting this delicious bite, holding the flavour for just a little while in your mouth.

Breathe in, breathe out.

Having filled your senses with this beautiful place, finding serenity in the calmness that nature offers you, take another breath, and gently breathe in the peace and richness, and exhale your gratitude for this safe space. Feel the strength in the ground beneath you. Slowly breathe in gratitude to yourself for taking this time to connect with nature, and, likewise, notice your sheer gratitude for the natural world. Feel complete oneness with nature.

Take another breath, gently, in and out. And when you're ready, slowly adjust your body, maybe stretching out.

Gently open your eyes and bring your awareness back into the room.

Chapter 8

From Anxiety to Action

Sustainable action

We're mindful that although action can alleviate anxiety, it must go hand in hand with self-care. We can think of this as *sustainable action*. By this, we mean action that is sustainable, for you, in the long term, as well as holding climate sustainability in mind. The CPA also notes that hyperactivity in activism can be counterproductive to maintaining wellbeing.[1] We should therefore remain mindful of the proportion of our time and mental energy we are dedicating to climate action, or engagement in whatever form (this includes, for example, researching or climate-related social media content). It is important that we don't let the meaning of these actions become what the CPA describes as a 'defensive response' that we use to compensate for, or mask, the emotional distress we feel. Three-month bursts of climate action followed by six months of climate depression just aren't that helpful from a pragmatic point of view, or a psychological one. If we balance our routine with self-care, connection, mental breaks and climate action, we will have a much more viable, authentic and sustained impact.

--- **VOICES** ---

When my climate anxiety was particularly bad, I would be signed up for multiple groups and working several jobs and picking up new habits, all in the name of fighting climate change. My climate anxiety usually made me feel like I wasn't doing enough and since we are running out of time, any time spent doing anything else was useless. It got to a point where relaxing or trying to pursue my hobbies made me MORE anxious, because I wasn't working on climate work. This always led to me burning out and quitting everything, which in turn kickstarted my climate anxiety again, and the cycle continued.

Former client (K)

What action should I take?

It's a jungle out there. How do we disentangle objective truths about the carbon footprint of certain activities from our own cognitive dissonance telling us that our actions are not that impactful, or that the biggest personal sacrifice probably isn't worth making? We propose three areas to consider, in order to begin (or continue) your journey to sustainable action. A sort of triangulation between what we suggest is the degree of *impact*, *meaning* and *ease* associated with each action idea. In a perfect situation, you are aiming, of course, for the centre of the Venn diagram (although, in many ways, you may well find that this area includes those actions that, rating high across all three criteria, you are already doing). Although the elegant simplicity of the diagram may not transcribe perfectly into your daily life, it is a worthy starting point in considering your next environmental move.

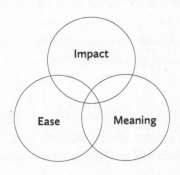

Impact

We'll be the first to admit that this is not our area of expertise. As psychologists, we are much more comfortable with helping people overcome the obstacles to action or supporting resilience in action than we are with guiding and advising people on which actions to take, from an objective standpoint. We are often asked, 'What can I do to have the biggest impact?' It's easy to become lost in answering this question, as the myriad variables, confounders and complexities of how our daily decisions impact the planet can quickly become overwhelming. Better to not act at all than take the wrong action? Inertia is a lurking predator here. This is an important question that we feel lands outside our remit. However, we would be remiss not to include some measurement of impact. Quite simply, we encourage you swiftly (while preserving our professional dignity) in the direction of people like Professor Mark Maslin. In his excellent book *How to Save the Planet: The Facts*, he provides the 15 most effective, evidence-based actions you can take, at an individual level, to have the greatest impact. In no particular order, he proposes you can:

1. talk about climate change

2. switch to a more vegetable-based diet

3. switch to a renewable energy supplier

4. make your home more energy efficient

5. use cars less

6. stop flying

7. divest your pension from fossil fuels

8. divest your investments from fossil fuels

9. refuse/reject excessive consumption

10. reduce what you use

11. reuse as much as you can

12. recycle as much as you can

13. use your consumer choice

14. protest

15. vote.[2]

That person within us who craves simplicity and order, of course, demands some form of ranking here. A league table of environmental actions. But this is next to impossible because it so very much depends on how your life is structured, and the extent to which you engage in each of these actions already. As a starting point, though, hold these 15 ideas in mind. Likewise, you can also become more Carbon Literate (with the Carbon Literacy Project)[3] or use one of a growing number of mobile apps to track your carbon footprint. There is also an increasing body of workbook literature on how to take climate action, such as *The Carbon Buddy Manual* by Colin Hastings,[4] or, for kids, the wonderful *101 Small Ways to Change the World* by Aubre Andrus[5] (which we particularly love as it combines activities to care for yourself, others and the planet).

Ease

No, really, it's okay. We give you permission to include feasibility in your decision-making. Be guided by the principle of least effort. The path of least resistance. We are much more likely to create longer-lasting change if doing so can be achieved with less personal upheaval. Remember the power of self-efficacy. If your expectations don't match your achievements, it will hurt.

'Ease' is personal to you and your family. You may already be eating a broadly (or exclusively) plant-based diet. You may work, and holiday, more locally. Your capital investments may already be sitting in ethical institutions. In which case, you're already doing a great job! Some of these actions, of course, may require greater upheaval. A further, less proximal shift from your current way of life. Also consider your Signature Strengths. By leaning into what you are already good at, you may well find the associated actions that much easier to operationalize. If 'Hope' was an essential strength, perhaps you have a role to play in sharing positive stories in your channels or motivating change in others. Likewise, 'Teamwork' might involve creating or supporting collective activism in your area. Psychologist Renée Lertzman has founded a fantastic online resource in this regard, Project InsideOut,[6] which tailors your individual skillset with related ideas and workshops for supporting collective climate action.

Jen Gale, author of the *Sustainable(ish) Guides* (to both Living[7] and Parenting[8]), has sage words regarding the importance of action that is accessible to you, by:

> making the best choices you can that work for you, your family, and the planet; accepting that we all start from different places, have different challenges and circumstances; no challenge is too small; aiming for progress not perfection; focusing on the things we can change, and not the things we can't.[9]

So start with what you are doing already, or what change feels easiest to first implement, and allow change to snowball(ish) from there.

Meaning

Much of this book relates to the importance of personal meaning in climate action. The mere fact that you have read through to Chapter 8 is itself a strong indicator of your Green Identity. You would be forgiven for thinking, at this stage, 'It all matters!' but return for a moment to earlier exercises. Remember your objects of care. How can these translate to actions?

For Patrick, meaning has often centred largely around two areas: wildlife conservation and renewable energy. So, for example, he often finds himself engaged in efforts to switch his networks – schools, hospitals, apartment buildings, neighbours – over to green energy suppliers. If something has high levels of personal meaning and resonance, scaling action upwards and outwards will feel both natural and rewarding.

Below we've included a space for you to record some possible actions, which score highly for impact, ease and meaning in your world. Are there any overlaps or areas of convergence? These may be great places to start.

Exercise: Acting with impact, ease and meaning

Possible impact actions

1. . . . .

2. . . . .

3. . . . .

Possible ease actions

1. . . . .

2. . . . .

3. . . . .

Possible meaning actions

1. . . . .

2. . . . .

3. . . . .

Motivating change in others

Motivating others is a fantastic way to stay motivated ourselves. Making shared commitments with others will improve our own accountability in achieving them, while also maximizing our impact. Here we examine a little social psychology before utilizing the power of nudge economics to help consider good, evidence-based ways of encouraging change in others.

The power of imitation is strong. We are hardwired from birth to imitate those around us – older siblings, parents and, later, peers. Through a sort of osmosis, we inherently understand and learn to imitate the social conduct of the world around us. However, it isn't just the behaviour of others. From a very early age, we mimic the gestures, posture, facial expressions, accent and all kinds of other mannerisms and characteristics of those around us. We do so without realizing, and this continues into adulthood. The wonderful thing about imitation is that its positive effects are two-directional. Not only do we imitate the positive behaviours of others, but when we are being imitated by others, it reinforces our own positive behaviours too. So, for example, although you may not realize it, you are more likely to, say, start home composting, if a sufficient number of your neighbours or friends do. But once they realize they have been imitated by you, they are likely to double down, increase their production or find further ways to avoid landfill. The same could be said for activism, political engagement, or any wider climate actions.

People sometimes avoid letting others know that they were the inspiration for their own behaviour. Somehow it can feel awkward, 'biting their style'. But imitation truly is the sincerest form of flattery, and so we encourage you to let people know. Have their climate behaviours inspired you? Tell them! Not only will it likely make them feel good, and reinforce their own self- and collective efficacy, but it will also likely lead to them furthering their own efforts with renewed vigour and enthusiasm.

Harnessing the power of 'nudge' economics

A nudge can be described as 'any aspect of the choice architecture that alters people's behavior in a predictable way without forbidding any options or significantly changing their economic incentives'.[10] There are a few key principles of behavioural economics that are central to a good 'nudge': 'libertarian paternalism', 'choice architecture' and 'default options'. The former refers, if you like, to the theory behind nudging, and the latter two, the practice.

Libertarian paternalism refers to the fundamental underpinnings of nudge economics – that is to say, its ideological basis. This is not about telling people what they should or shouldn't, must or mustn't do. Instead, people should have the option to decide themselves with minimal control (libertarian), but be guided and supported in their decision-making process to make the right decision, for themselves and indeed the planet (paternalism). Nudges 'care' about the person and environment. They are concerned for the best short- and long-term outcomes. Libertarian paternalism therefore remains palatable to many a political persuasion.

As Thaler and Sunstein describe, 'Just as no building lacks an architecture, so no choice lacks a context'. Choice architecture essentially describes the way a person's options are crafted and presented to them. A choice architect, therefore, is the agent responsible for the presentation of these decisions (i.e. you!). And they are quick to reassure that choice architects can come from all walks of life. Indeed, every single person, in some form or another in their daily lives, acts as the architect for the decisions of others:

> If you are a doctor and must describe the alternative treatments available to a patient, you are a choice architect. If you design the forms that new employees fill out to enroll in the company health care plan, you are a choice architect. If you are a parent, describing potential educational options to your son or daughter, you are a choice architect. If you are a sales person, you are a choice architect (but you already knew that).[11]

Related to the concept of choice architecture is the knowledge that no choice is made in isolation. That is to say, humans are social animals, and therefore our behaviour is more likely to be influenced by whether it is perceived to be desirable to others in our communities (the 'social desirability bias') than it is by facts and figures alone.

Default options are among the most commonly used nudge interventions and have worked to good effect in supporting people to make pro-environmental decisions. People are quite simply more likely to choose the default among a series of options. 'Would you like to receive additional marketing from us?'

Nudging became such a potentially potent tool across the world that governments established their own 'Nudge Units', the first being the Behavioural Insights Team, formed in the UK in 2010. This was followed by similar units in the US, Canada, Germany and Australia, while a number of other countries also began using the insights from nudge interventions to form policy decisions, including India, Singapore, Peru and Indonesia.[12] Indeed, according to one landmark report, *Nudging all over the World*, 136 countries/states had observed 'nudge'-style approaches having some sort of effect on public policy.[13] In fact, this report challenged many preconceived notions that nudges were solely the interest of developed countries, highlighting the incredible and widespread work of nudging in less economically developed countries, particularly in the areas of public health, including preventing the spread of HIV/AIDS and malaria.

Although many classic nudges focus on encouraging people to make better long-term decisions, or, at least, choices are presented in such a way as to make these the most likely option, in areas such as health and personal finance, increasingly nudges became concerned with pro-environmental behaviour. These became known as Green Nudges.[14]

Are nudges ethical?

Often arguments about the ethics of 'nudging' hold the assumption that humans are effective decision-makers, especially regarding choices with implications on long-term outcomes. However, as much of this book has already outlined, although humans hold an immense capacity, unparalleled in the natural world, to weigh evidence, rationalize, plan and decision-make, ultimately we often make unconscious, System 1, easy decisions. Therefore, the argument for nudging becomes stronger. Sometimes we act in spite of ourselves.

If we are indeed primed to favour the status quo, is there not a moral obligation (not least an opportunity) to design the presentation of choices in such a way as to protect the future person and planet? In many ways, provided the guidance and framework of nudging (as opposed to restricting choice) is followed, would it not be ethically irresponsible not to?

Some classic 'Green Nudges'

Nudging at the table

What if we could change people's eating habits, not by telling them what they are and aren't allowed, but by presenting their options slightly differently? This is precisely what a number of studies have attempted to do, and often with good effect.

Availability and access are often manipulated here. For example, putting less healthy foods in the centre of the salad bar (where it's harder to reach) can lead to less uptake of those foods. Likewise, putting two serving forks on the tray of the preferential food, so it can be accessed from both sides by two people, increases uptake.[15]

Researchers experimenting with the size of plates being given to guests at a hotel found that food waste was reduced by around 20%. Importantly as well, there were no changes found in guest satisfaction, leading to what the researchers described as a 'win–win'.[16] This is a great example of the default option. Here, there were no limits made as to the amount of food guests were entitled to, but merely the size of their plate. This has the potential to be adopted in a widespread manner, from the dinner table to wherever your mind takes you. However, it is important to highlight that the 'smaller plate' effect has had mixed results in the literature,[17, 18] suggesting that perhaps a much more reliable means of reducing food intake (and therefore waste) is to control portion sizes.[19]

The 'follow the crowd' effect

Capitalizing on people's inherent and automatic drive to imitate those around them is argued to be the most effective tool in the nudging belt, and probably the most easily accessible to the 'everyday' choice architect, at work or in their community.

US software company OPower worked with energy companies and began sharing data with customers not only regarding their own energy use but also how it compares to that of their neighbours. Analysis showed that, on average, people's energy usage went down by 2–3%.[20] Although, at an individual household level, this may not seem like much, it has huge scalability with a minimal cost and, again, without limiting people's individual freedoms. This can be seen as demonstrating the power of socially desirable behaviour.

Indeed, energy usage has been the primary driver of Green Nudging at an organizational or national level. In Germany's Black Forest, the default option approach was taken to good effect by energy companies, by switching customers' default choice to renewable energy sources rather than fossil fuels (of course, they had the right to 'opt out' at any time, thus not limiting the choices available in any way).[21]

You have to be careful with the 'Follow the crowd' effect, though. Some studies reveal that, actually, by telling a particularly eco-conscious or low-energy-using family that they are consuming less than their neighbours, you can inadvertently make using *more* energy seem like the most socially desirable thing to do. This can be prevented by, for example, including the image of a smiling face to signify a good energy score and a frowning face to indicate bad. This can have the effect of making clear the fact that less consumption is indeed the socially desirable behaviour.

Nudges at work

Did you know that when departments in Rutgers University changed the (default) option on their printers to 'double-sided', it saved as much as 620 trees in paper over a single semester? This was replicated in a Swedish university, which reported a 15% reduction in daily paper usage.[22] This, it is suggested, works better than applying a penalty on paper use. Think of all the printers in all the offices in the world. This simple nudge could, when scaled, make a huge difference.

Other popular nudges in the workplace have been aimed at making pro-environmental behaviour more fun. These might include footprints painted on the stairs (and away from the elevator), basketball hoops above the recycle bin, or even targets drawn on urinals at the precise point that eliminates mess!

Exercise: Becoming a choice architect

Well, actually, the title of this exercise is misleading. You are already a choice architect. In some form or another, each and every day, you have the potential to engineer the choices of the people around you. Any moment in the day that involves either direct or indirect contact with somebody else, or where your action in some way determines the options available to another person, is a potential nudge moment. Think about opportunities to set green default options or social desirability. Record some ideas below.

Nudges at home:

. . . .

Nudges at work:

. . . .

Nudges with friends:

. . . .

Nudges in my wider community:

. . . .

Any other nudges:

. . . .

The Ethical Nudge project

We created an Instagram handle (@ethicalnudge), where we share ideas, some our own, others from the research literature, real-world experiments, and other recommendations from our followers. We invite you to join and share your own nudges!

The (potential) pitfalls of individual action

Generally, although we advocate for moving from anxiety to action, it doesn't necessarily translate that action is the *cure* for anxiety. Activism, it is said, can tame difficult emotions[23] but does not alleviate them. Nor should it. Many of the people we speak with are engaged in climate action already and are nonetheless feeling the effects of anxiety, grief and rage. There should not be, therefore, an unfair expectation that taking individual action will inoculate against any future psychological ill-effects of climate change.

There is some suggestion that an over-focus on individual actions (recycling, driving an electric car, buying local, organic produce) can come at the expense of collective political participation. This can be problematic for several reasons. First, collective action works. Quite simply, action that affects policy decision-making can create the sort of shift required to properly tackle global warming at a systemic level. Also, an over-focus on individual actions can leave you feeling isolated in your efforts, especially if you are not seeing similar action in your community.

All this pitfall requires is mindful attention being paid to the type of action you are engaging in. Try where possible to find a balance of climate actions around the home, the workplace and, in some way, at a wider level as well – keeping the action sustainable for you, of course.

Then there is the potential for either responsibility or despondency to impede action, by noticing all of your fantastic individual accomplishments, but not seeing them replicated by those around you, or, similarly, that creeping feeling that despite your best efforts, the problem of global warming remains. In these scenarios, familiar to many an environmentally minded person, one faces the risk of a potent emotional backlash, either by increasing a sense of individual responsibility and leading the person to feel the need to double down on their efforts, or instead an almost opposite reaction of dejection ('What's the point?').

In both of these emotional scenarios (responsibility and dejection), it might help to hold in mind the fact that our individual actions *do* affect the behaviours of others. Remember the power of imitation, of social norms and desirability. Remember the OPower effect (of sharing the energy use of their customer's neighbours). Although you may not be able to directly see how your actions affect the choices of others (or maybe you can), the power of social psychology is on your side.

Perfect gets nothing done

with perfectionism into this imperfect organic flow of emotion and work that blossoms into reality. That's where activism can thrive.

— Isaias Hernandez (@QueerBrownVegan) – environmental educator

Our daughter came skipping home from school one day. We asked, 'Did you have a perfect day?' She responded, 'Perfect gets nothing done. Practice makes progress!' (We couldn't agree more. Also, can we give her teacher a raise?!)

Perfectionism is one of the pitfalls of climate action. This moves us back to that position of feeling that this is down to us to solve. That we aren't good enough. We will again always point out where we think people (though it's usually more often systems) could be challenged to participate more, to do better. Acknowledgment that we are part of a system that is imperfect will refocus the self-blame back on the responsible parties – big businesses and governments – that perpetuate the crisis. Perfectionism can also deter our ability to take action, diminishing our self-efficacy.

There will be times when our objectives are unsuccessful. How we experience and cope with these setbacks will shape both our mental wellbeing and our impact on the planet. Jen Gale highlights the fallibility of our perception that we need to do climate work all the time, perfectly. Rather it is about:

> Doing what you can, one baby step at a time. No preaching, no judgments, no expectation of 'eco-perfection'. I got fed up with the narrative that we can only make a difference if we live off-grid in a yurt and learn to knit our own yoghurt. We all have a huge potential to create positive change, imperfectly.[24]

Although often we can be inspired by the stories of those who are able to live off-grid or work full-time in activism or develop pro-climate technologies, for many that is not an option. We can certainly make choices in how we spend, what we consume and how we participate, but there is no such thing as the perfect climate citizen. There is a narrative of perfectionism in our culture, personified by social media, which can be debilitating and has a huge impact on mental wellbeing. The myth of perfection can lead us to avoid the necessary action.

Scheduling is self-care

--- **VOICES** ---

I schedule breaks throughout the day. I need the schedule. I kept trying to meditate but I'd fall asleep. I have a fish tank and I found myself just staring at my fish one day and I realized – this is actually me taking a break. So I changed my schedule from 'meditation' to '30 minutes stare at your fish!' I really go out of my way to know that's part of my day. I get so much satisfaction from checking things off my schedule. I don't have to think about it in my head. Getting to check it off, even if it's 'taking a break'. I think of this as our duty as an activist. I know this is so easy for me to forget to take a break so I make sure I do that.

— *Mitzi Jonelle Tan, climate justice activist from the Philippines* —

For a book about climate anxiety, you might be wondering why we are so emphatic about scheduling.[25] It's because we want everyone to be as productive as possible to mobilize against climate change. But the practice of scheduling is not only *productive*; it is also *protective*. Just as children value containment, so do adults, especially when we are struggling with our experiences of big climate emotions. If we know *which* actions are expected of us (or perhaps which actions we expect from ourselves), if we are clear on *how* we can get them done, and *when* we will commit to completing these, they simply become tasks that we cross through as they are achieved.

We don't invest emotionally. Rather than thinking about what we *should* do, and feeling guilty or annoyed with ourselves for not doing it, we just do it. That might seem obvious, but we are shocked at how often we revert to that old, reactive part of our brain, rather than tapping into the pragmatic and evaluative, which encourages us to structure for more efficient and well-thought-out results. The implementation of scheduling has helped our clients achieve balance. It's one piece of feedback that we consistently hear: scheduling really helps. We want organization and methodology! Scheduling not only provides this but also allows us to recognize our efforts and celebrate our successes.

When coaching, one of the first things we want to bring into focus is the client's feelings of self-worth in relation to how they value their time and goals. This helps to reveal the mental obstacles that are deterring them from success. We cannot give meaningful time to anyone else fully without truly valuing our own time. Remember, when we feel scarce, time can feel insufficient, but there is time for everything if we manage it well. By valuing our time, we value our work, and thus are more productive and effective in what we create. Whether tackling anxiety or depression, starting a new business or coordinating carpools, we can see the benefits of evaluating how we spend our time; the importance of a given task. We commit to or eliminate jobs, chores or ventures that are unnecessary, that do not bring value. We show up when it is required. Climate change requires us all to take action and it is fundamental for us to do this in a strategic and deliberate way.

Hitting hard reset

Sometimes taking a breather is what you need to get back into the game. As research professor and resilience guru Brené Brown reminds us, 'When we hit that wall, sometimes courage looks like scaling it or breaking through it. And sometimes courage is building a fort against the wall and taking a nap.'[26] We call this 'hitting hard reset'. View this reset as a necessary update to your mental software. We are debugging, slowing down so that we can be more efficient, deleting messages that we don't need, extending our bandwidth and seeing less of the buffering pinwheel of doom.

Making the most of this 'breather' will help you gain more strength to return with clear intentions and a lighter mind. This is particularly important if your profession is climate-related, and thus much of your day is spent engaging with the problem.

How long this reset requires will vary, of course, but a helpful starting point is to take a weekend. Planning this will provide a level of containment, allow you to feel in control and protect against free time that could result in rumination, doomscrolling or other depleting behaviour. What's important over this period is that you notice if you are becoming anxious, alert or perhaps feeling guilty and gently guide yourself back to the necessity of reflection and recuperation. One initial hurdle to overcome, as many have told us, is *giving yourself permission*. Start the weekend reset with a reminder to yourself that you are allowed this time. It is essential maintenance.

Actively plan out your weekend. We encourage you to schedule this. For example, morning exercise and meditation. Start as you intend to go on. Evening meal preparation with a friend for dinner. Schedule as much outdoor time as you can manage. It is important to reconnect with the nature that you love. If you've spent all week trying to nurture the planet, allow yourself to be recharged by its beauty and calm. Schedule in time to connect with others and time to be on your own. This is not a weekend to engage with social media, news or climate content. It will be there on Monday. Commit to minimizing your technology time. If you watch a film or TV show, fine, but refrain from *A Life on Our Planet* or *Seaspiracy*, powerful as they are, tempting though it may be. At least for this weekend.

We encourage you to end this period of time by writing down what you have felt, what thoughts have come to you and what ideas have emerged. The hard reset is a time to enjoy space to become creative. It may well be that you feel inspired by new projects, new ideas. Although that may be a beneficial *outcome* of the hard reset, it is not the *intention*. This time is about recuperation, not finding new solutions. If there is a constant stream of noise, negative energy and anxiety, where is the room for growth? At the end of this hard reset, schedule for the week ahead.

Strategic goal setting

Similar to our stepping stones exercise, imagine our goal is sitting at the top of a ladder. We can't just leap up to the top. We have to climb, step by step, to where we want to be. If we slip, we catch our footing or take a hand from a friend and continue our climb.

Return to your 'Gauging your whole-life balance' exercise in Chapter 7. Let's experiment with setting a goal to increase one of the areas in which you felt you were struggling. For example, if in your health arena you felt you needed to pay more attention, you might think, 'I should be more physically active.' Often people think this is a *goal* when actually it is a *thought*. If we set a goal for ourselves (at the top of our ladder), we can then define what each step would be in order to reach our goal. If my *goal* is to run a 10K in a month's time, then I need to think strategically about what I need to make that happen. Essentially, start at the top and work your way back down.

'I should work out more' is not specific enough. We can shift this to 'I will run three days a week.' Then shift to 'I run on Monday, Wednesday and Saturday mornings for 45 minutes while I listen to my favourite podcast.' I now stand a much better chance of achieving my goal because I have a scheduled plan.

We can be not only strategic when we organize in the realm of our climate contributions, but creative as well. Our current thinking might be 'I feel like I am filling in lots of petitions about plastic waste, cutting back at home and writing letters but it's not enough'. Instead, we can think about what goals we want to achieve and what steps we need to take to make them happen. 'I want to clean up all the plastic' can shift to specific plans: 'My goal is to raise $1000. Then I will donate the money to 4Ocean who will remove plastic from the sea' – at the time of writing, totalling 15.7 million pounds in weight![27] Then we can work backwards to make a real plan: 'I am going to run 100 miles over a month and use my networks to get friends to sponsor me. I will set up the GoFundMe page today. I will post my runs on social media and challenge other people to join me. I will strategically hashtag. I will start running on the first of next month.'

Do we see how many boxes we can tick if we have the mental space to get creative? Notice, we are committing to a time frame, and to others about our intentions. By doing this we are much more likely to achieve success. We are also combining something that we enjoy (if running appeals to you, that is) and that is healthy for us. When we get creative, we can structure goals that tick more than one box.

Exercise: Setting strategic goals

When we make plans to implement our goals, we should ask ourselves:

Is this a thought or a goal? (If not already, turn into a specific, achievable goal.)

. . . .

. . . .

What actions do I need to take to achieve this (working down the ladder)?

. . . .

. . . .

What is my timeline?

. . . .

. . . .

Who can I tell? Can I get people to join me?

. . . .

. . . .

How can I maximize its wider impact?

. . . .

. . . .

How can I make this enjoyable?

. . . .

. . . .

How will I celebrate my success?

. . . .

. . . .

Values, goals and actions

We talked about the benefits of identifying goals that resonate with your values. Here we can translate those goals into actions and include them in our schedule. This gives us the structure we need to actualize them. We can separate these by the *why* (values), *what* (goals) and *how* (actions) using the guide on the next page.

Exercise: Translating values into actions

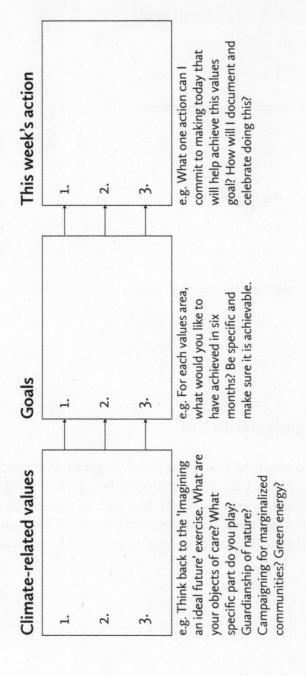

Climate-related values

1.

2.

3.

e.g. Think back to the 'Imagining an ideal future' exercise. What are your objects of care? What specific part do you play? Guardianship of nature? Campaigning for marginalized communities? Green energy?

Goals

1.

2.

3.

e.g. For each values area, what would you like to have achieved in six months? Be specific and make sure it is achievable.

This week's action

1.

2.

3.

e.g. What one action can I commit to making today that will help achieve this values goal? How will I document and celebrate doing this?

Emotional and mental real estate

As Mark Twain instructed, 'Eat a live frog first thing in the morning and nothing worse will happen to you the rest of the day.' Don't eat frogs – we love those little guys. But there is something to be said for getting the worst out of the way, nice and early. By dwelling on, or avoiding, all the things that rack up time and energy in our minds, all the things we *should* or *need* to do, we end up causing ourselves more stress and discomfort than if we had just sucked it up and got on with it. This energy takes up our *emotional and mental real estate* and it's exhausting. Sure, there will be the inevitable life admin requirements, filing taxes, running errands – these are things that have to happen – but when we start to view time as an honoured rather than a scarce commodity, we are more prudent and deliberate in how we use our time. Rather than what we *should* do, we simply do. We then waste less time resisting and 'eat our frogs', inherently freeing up that prime real estate for the actions we care about.

By scheduling in those proverbial frogs, we know that we can learn to commit to our time and decisions in a way that serves us. For example, 'I know that the first thing I am going to do at 9am this morning is deal with those bills. They will take 45 minutes. When they are complete, I can begin my emails for the climate event I am organizing. I won't sleep late or keep getting up to make cups of tea or looking at my WhatsApp. Even though I think bills are boring, I will complete them because I am committed to the value of my time and that can support my climate work.' By guarding this mental and emotional real estate, we can refuse temptations like procrastination or distraction in order to maximize our time spent doing the things that we enjoy and care about.

Consciously scheduling

For some people, lists are effective. However, we often find that lists can feel vague, without concrete structure or strategy. Lists are vague, without concrete structure or strategy. Transfer your dependency on list-making, or keeping things in your head, on to a physical schedule instead. When you start out, you want to take your lists (whether mental or Post-it notes littering your desk) and compile these into one place. Decide how long each task on your list will take. Be mindful not to give yourself too much, or too little, time to complete your tasks. Make them realistic but also don't allow enough time to dawdle or become distracted. The next time you remember something you need to do, rather than adding to a list, schedule it in. Free up the mental real estate.

Return to your 'whole-life balance' activity. Which areas were registering low? Incorporate activities that will boost these areas. Don't overlook the importance of self-care and include what nourishes you. First, put in the time that you need to strengthen your physical and mental wellbeing. This will motivate you to get all the frogs out of the way so that you can do what you enjoy.

As we develop this schedule, hold your climate goals in mind. Think strategically about where you feel most resilient in the week to engage with these tasks and bookend with something restorative.

Then schedule, schedule, schedule.

An example of a schedule is below. It is merely an example that doesn't include a role as a parent or carer. It doesn't assume you are also working full-time in a climate-related profession. For some, this will include more items like 'Family litter picks', and for others, 'Social media black-outs'. There may be a need to structure in 'contingency time', whereby you leave more room to buffer for things that are beyond your control, like an unpredictable working schedule.

	Mon	Tues	Wed	Thurs	Fri	Sat	Sun
Morning What will you do? Where will you do it? Who will you do it with?	Wake up 7 Breakfast/read news 15 mins Mindfulness 10 mins Walk to work	Wake up 7 Breakfast/ social media time 15 mins Yoga YouTube 15 mins Bus to work – sign petitions on phone during journey	Wake up 7 Breakfast 15 mins Mindfulness 10 mins Walk to work	Wake up 7 Breakfast/read news/social media 15 mins Mindfulness, 10 mins Walk to work	Breakfast/ social media time 15 mins Yoga YouTube 15 mins Bus to work – sign petitions on phone during journey	Lie in until 9am but no later Mindfulness 20 mins Call Mum for a catch-up Read 30 mins (first 15 can be climate-related)	Get up at 8am Mindfulness 20 mins Tidy house 30 mins Bike ride and coffee with Justin Tidy house 30 mins
Afternoon What will you do? Where will you do it? Who will you do it with?	Lunch with Sarah in park 45 mins Email for Fridays for Future organizers and speakers 20 mins	30-min walk Follow-up emails 15 mins	Round table Zoom for Fashion Revolution	1-hr walk with Holly and quick lunch	Lunch listening to podcast 30 mins Walk outside 20 mins	Research some climate clubs to join and sign up 45 mins Go for a long walk 2 hrs	Lunch with Paul Review my last week Social media or reading Scheduling the week 20 mins
Evening What will you do? Where will you do it? Who will you do it with?	Gym 1 hr Dinner Reading Bed by 10	Laundry and cleaning 1 hr Watch a film Bed by 10	Food shop Dinner Reading Bed by 10	Panel talk 45 mins Mindfulness 10 mins Dinner	Call Dad Dinner in with Marianna and Katixa	Dinner out with Georgie to celebrate round table organization	Watch a film In bed by 10

This is not prescriptive but meant to give an idea of what a balanced week can look like. Tasks and timing for activities are contained. We invite you to try this now. Committing to action is something that we know is difficult, especially when we might be feeling that our self-efficacy is low or 'It's just one more thing to do'. A key component of coaching is requesting commitment. We would request that you commit to honouring your schedule for two weeks. Clear the decks on those frogs, show up for your physical and mental wellbeing, and structure your plans for climate action strategically. You may experience some setbacks (see below); work through them but stay committed to trying this new system. Tell a friend you are going for it. Organize for yourself, organize for the planet.

	Mon	Tues	Wed	Thurs	Fri	Sat	Sun
Morning What will you do? Where will you do it? Who will you do it with?							
Afternoon What will you do? Where will you do it? Who will you do it with?							
Evening What will you do? Where will you do it? Who will you do it with?							

Signal and celebrate

Megan met an Extinction Rebellion climate activist during the 2019 Autumn Rising. This person had been arrested numerous times, had given countless speeches and organized events. Megan was curious and asked, 'What are you going to do after this?' They replied, 'Now it's the time for some quiet and some reflection and to just take this all in. We did good work here. I definitely need to be in nature for a while; knowing I have that next feels good.' This person was a great example of someone deeply committed. Impactful, yet aware of the importance of acknowledging their work. Knowing how essential it is to recharge, to 'bookend' climate activity and to celebrate.

Imagine looking back over your schedule at the end of the week. If you crossed through a few of the items, it doesn't relay the sense that you have achieved much. It is very likely that the six items you crossed off are, in reality, only a fraction of all the other tasks you completed. But looking back, you don't gain that sense of accomplishment. After integrating scheduling, along with signalling, one of our clients realized they were actually signing close to 40 climate petitions a week, along with all their day-to-day activity. Once they had the chance to record this activity, to see it written before them, it became apparent that they were achieving more than they had initially predicted. Through this, they gained a renewed sense of motivation to bring to their climate work.

Likewise, scheduling can also help hold yourself accountable for those unfinished tasks. Without judgment, ask yourself *why* certain tasks don't get crossed off. Are you seeing a pattern when it comes to missing, for example, your self-care targets? Are you procrastinating in your climate work, perhaps because you are finding it more emotionally challenging than usual? Or, perhaps, simply not all of these actions made it on to the schedule. If so, what can we do to support ourselves so that we can overcome the obstacle and get on with what we set out to do? Intentionally 'ticking off' signifies progress, which promotes self-efficacy.

	Mon	Tues	Wed	Thurs	Fri	Sat	Sun
Morning What will you do? Where will you do it? Who will you do it with?	Wake up 7 Breakfast/read news 15 mins Mindfulness 10 mins Walk to work	Wake up 7 Breakfast/ social media time 15 mins ~~Yoga YouTube 15 mins~~ Bus to work – sign petitions on phone during journey	Wake up 7 Breakfast 15 mins Mindfulness 10 mins Walk to work	Wake up 7 Breakfast/read news/ social media 15 mins Mindfulness 10 mins Walk to work	Breakfast/ social media time 15 mins Yoga YouTube 15 mins Bus to work – sign petitions on phone during journey	Lie in until 9am but no later Mindfulness 20 mins ~~Call Mum for a catch-up~~ Read 30 mins (first 15 can be climate-related)	Get up at 8am Mindfulness 20 mins Tidy house 30 mins Bike ride and coffee with Justin Tidy house 30 mins
Afternoon What will you do? Where will you do it? Who will you do it with?	Lunch with Sarah in park 45 min ~~Email for Friday for Future organizers and speakers 20 mins~~	30-min walk Follow-up emails 15 mins	~~Round-table Zoom for Fashion Revolution~~	1-hr walk with Holly and quick lunch	Lunch listening to podcast 30 mins Walk outside 20 mins	Research some climate clubs to join and sign up 45 minute Go for a long walk 2 hrs	Lunch with Paul Review my last week Social media or reading Scheduling the week 20 min
Evening What will you do? Where will you do it? Who will you do it with?	~~Gym 1 hour~~ Dinner Reading Bed by 10	Laundry and cleaning 1 hr Watch a film Bed by 10	Food shop Dinner Reading Bed by 10	Panel talk 45 mins Mindfulness 10 mins Dinner	Call Dad ~~Dinner in with Marianna and Katixa~~	Dinner out with Georgie to celebrate round table organization	Watch a film In bed by 10

Now imagine looking back and seeing a week where we took the time to signal what we had accomplished, perhaps even adding in the other activities or work that we manage to slip in on top of our tasks, self-care, community connections and goals.

	Mon	Tues	Wed	Thurs	Fri	Sat	Sun
Morning What will you do?	Wake up 7 Breakfast/read news 15 mins Mindfulness 10 mins Walk to work	Wake up 7 Breakfast/ social media time 15 min Yoga YouTube 15 mins Bus to work—sign petitions on phone during journey	Wake up 7 Breakfast 15 mins Mindfulness 10 mins Walk to work	Wake up 7 Breakfast/read news/social media 15 minutes Mindfulness, 10 minutes Walk to work	Breakfast/ social media time 15 mins Yoga YouTube 15 mins Bus to work—sign petitions on phone during journey	Lie in until 9am but no later Mindfulness 20 mins Call Mum for a catch-up Read 30 mins (first 15 can be climate-related)	Get up at 8am Mindfulness 20 mins Tidy house 30 mins Bike ride and coffee with Justin Tidy house 30 mins
Where will you do it?							
Who will you do it with?							
Afternoon What will you do?	Lunch with Sarah in park 45 mins Email for Fridays for Future organizers and speakers 20 mins	30-min walk Follow-up emails 15 mins	Round table Zoom for Fashion Revolution	1-hr walk with Holly and quick lunch	Lunch listening to podcast 30 mins Walk outside 20 mins	Research some climate clubs to join and sign up 45 minute Go for a long walk 2 hrs	Lunch with Paul Review my last week Social media or reading Scheduling the week 20 mins
Where will you do it?							
Who will you do it with?							
Evening What will you do?	Gym 1 hr Dinner Reading Bed by 10	Laundry and cleaning 1 hr Watch a film Bed by 10	Food shop Dinner Reading Bed by 10	Panel talk 45 mins Mindfulness 10 mins Dinner	Call Dad Dinner in with Marianna and Katixa	Dinner out with Georgie to celebrate round table organization	Watch a film in bed by 10
Where will you do it?							
Who will you do it with?							

Reminding yourself of your achievements isn't self-indulgent. This is harnessing the power of good behavioural science to your advantage. Without comparing you to toddlers or puppies, our brains respond well to rewards. We like to see a quick positive response. Global warming is too big for one of us to solve. We are playing a long game, which means it is all the more important to acknowledge our wins, big and small. This makes us feel abundant and motivated. Celebration affords us the chance to recognize the important role we have to play in climate action and will perpetuate our ability to continue.

Feeling okay to feel okay

This is an area where we have felt humbled and enriched by the feedback of those we have worked with. These are the nuances to recovery that can catch us, as psychologists and individuals, slightly off guard. In a sense, how can we allow ourselves to be okay with feeling less anxious, experiencing less guilt or rage? When the stimulus of these emotions is a problem that we are deeply committed to fixing for the planet, there can exist a secondary thinking process. Sometimes we call these 'meta-cognitions', whereby our thinking becomes less about the 'problem at hand'; instead, we gaze inwards: 'What does it mean about me that I am responding to the problem in this way?' Perhaps there may be a concern that, by managing the intense emotional experiences, it will mean you care less? Or will become complacent or less committed to action?

Remember, at times such as these, about the thinking trap of emotional reasoning. Am I viewing my *emotion* of guilt as evidence of a *fact* that I am doing something wrong? Or not doing enough? Remind yourself of the achievements you have had, the changes you made, and reach out to your networks. Talk, if you can, about your emotional experience. You may well find that others have felt the same struggle in learning to be okay with feeling okay.

VOICES

If you are successful at implementing a strategy to manage your anxiety and you see material improvement, you might end up with a feeling of guilt. I know I still sometimes do – when something triggers me and I handle it really well, I can be left with a lingering question: Shouldn't you be more worried about this? Is it okay for you to feel okay right now? The easiest way I've found to manage that feeling is to sort out my life and know that I'm taking every action I can possibly take as an individual to have an impact on the issue. Worrying won't change anything and so you shouldn't feel guilty about not doing it. Voting for politicians who understand the science and have a plan to resolve climate change, buying an electric car, offsetting or cancelling your long-distance flights in favour of trains or boat rides, and reducing the amount of meat in your diet will.

Former client (C)

Setbacks and how to respond to them

Having a fallback plan is a useful way of knowing how to get going when the going gets tough. Designing this before having a setback will lighten the blow and ground you so that you can process this setback and move forward. If you experience a setback in your climate work – perhaps a goal missed, or a project that didn't quite get off the ground, a system that said 'no', or a person remaining 'unnudged' – we propose a system of how to handle the fallout.

The SETBACK steps
S. Seek support. Who has your emotional back? Who can you confide in? Who will understand? Who will encourage you? Ask them

if you can contact them before you need them to establish this as a safe, non-judgmental lifeline.

E. Evaluate the setback. What didn't work and why? Can it still work? What, if so, can you do to move forward?

T. Take on an easy task. If you have just suffered a big setback, hit reset by engaging in one (or a few) simple tasks that support the planet. Try to keep this localized so you can see some direct results. Do a litter pick, donate to a charity, plant some bee-friendly flowers – anything that gets you back in the saddle.

B. Believe that you can overcome setbacks and move forward. Remind yourself of all the positive things you have accomplished – climate-related or not.

A. Acknowledge your responsibility. Did you show up for yourself? Did you honour your objectives? This also means not berating yourself if it was beyond your control. If you were not responsible for the setback, give yourself permission to release the guilt or 'what if I had just...' thinking.

C. Compartmentalize. Our setbacks do not define us unless we let them. We want to accept that setbacks are a normal part of life and don't mean that you are back at square one. One setback does not delete all the other positive actions you have achieved. Put this into perspective.

K. Keep moving forward. You've got this.

Exercise: My SETBACK plan

Here we ask you to map out your own SETBACK steps to help you stay resilient and dynamic when it doesn't quite work out the way you'd hoped.

S ....

E ....

T ....

B ....

A ....

C ....

K ....

There, now we have a plan. From this we can explore the possibility that it isn't necessarily a fail but a new way of figuring things
out. We are dynamic and resilient, and we will find a way to move
forward.

Staying well: your blueprint (for a greener planet)

Well, that is that. But it's not that, is it? You might be feeling all
charged up and ready to roll, but perhaps, in a week's time, you
might feel overwhelmed or disillusioned. What we hope you've
taken away is that that is okay. Climate work is a fluid space,
but you now have the tools to navigate it. We asked you to work
through the book sequentially in order to more deeply understand
your emotional relationship with the climate emergency. Now, we
invite you to dip in and out depending on what you need. Reach
for this book to remind you of your defences, to validate your
experiences, to motivate and organize yourself and to connect
with both nature and your community. We thought about writing
our *last chapter*, but we aren't here now to say goodbye. This is a
nest you can fly home to. As we hope you have learned, reach out
when you need it.

We hope that perhaps you have shone a light into those darker corners, illuminated some defences, thinking traps. Maybe you have made a commitment to one or more value-laden actions. Maybe you found, or became reunited with, an object of care. We sincerely hope that you feel grounded in your emotional response to climate change and that you have a sense of what sustainable action means for you, right now, in your life.

The 12 steps to turn the tide on climate anxiety

1. Understand and accept the facts.

2. Validate and normalize the psychological impacts of these facts.

3. Communicate productively and positively about facts, emotions and actions.

4. Form connections through stories.

5. Immerse yourself in nature.

6. Start with self-care.

7. Create your sustainable action plan based on impact, ease and meaning.

8. Take both individual and collective action.

9. Schedule your activity.

10. Allow for, and respond effectively to, setbacks.

11. Celebrate success.

12. Nudge the others.

Together, we are unstoppable

Science will save us but not without the stories that engage us.

We celebrate you. All of you. We are all part of this movement. Together. Thank you for all that you do.

We reflect on the opportunity we have had to learn from the experiences of our contributors. We are awe-struck by their kindness, humility, honesty and perseverance. They symbolize our hope of what we can all become. Imperfectly impactful. Mindful and meaningful. Responsible and resilient. Invested, impassioned and important on our beautiful, improbable earth. They represent all of you.

Al Gore said, with conviction, 'I believe the sustainability revolution is unstoppable.'[28]

So do we.

About the Authors

Megan Kennedy-Woodard is a coaching psychologist and Dr Patrick Kennedy-Williams is a clinical psychologist. They are Co-directors of Climate Psychologists, an independent organization providing individual therapeutic support and wider consultation regarding the mental health implications of climate change. Central to their work is climate change communication. They work with individuals, parents, educators, national government and media organizations to promote psychologically informed, constructive communication, to inspire hope and action in combatting climate change.

www.climatepsychologists.com

You can also follow the authors on Instagram (@climatepsychologists) and Twitter (@CPsychologists).

Endnotes and References

Chapter 1

1   Manson, M. (2016) *The Subtle Art of Not Giving a F*ck: A Counterintuitive Approach to Living a Good Life*. New York, NY: HarperCollins, p.15.

2   The APA, incidentally, have, increasingly over the past few years, been throwing their weight behind understanding the intersectionality of climate change and mental health.

3   Clayton, S., Manning, C., Krygsman, K. and Speiser, M. (2017) *Mental Health and Our Changing Climate: Impacts, Implications, and Guidance*. Washington, DC: American Psychological Association and ecoAmerica, p.4.

4   Ray, S.J. (2020) *A Field Guide to Climate Anxiety: How to Keep Your Cool on a Warming Planet*. Oakland, CA: University of California Press.

5   Ray, S.J. (2021, 21 March) 'Climate anxiety is an overwhelmingly white phenomenon.' *Scientific American*. Accessed on 15/06/2021 at www.scientificamerican.com/article/the-unbearable-whiteness -of-climate-anxiety.

6   White saviourism is a means of describing efforts taken by those in (white) privilege to aid, assist or 'rescue' non-white people in a self-serving way. It holds, with good reason, an increasing presence in the field of intersectional environmentalism and serves us, as authors of a book on climate anxiety, as an important blind spot, in terms of the messages we give and the stories and stereotypes we hope not to perpetuate.

7   Kazak, A.E. (1997) 'A contextual family/systems approach to pediatric psychology: Introduction to the special issue.' *Journal of Pediatric Psychology 22*, 2, 141–148.

8   Stoknes, P.E. (2015) *What We Think About When We Try Not to Think About Global Warming: Toward a New Psychology of Climate Action*. White River Junction, VT: Chelsea Green Publishing.

9   Climate Outreach (n.d.) 'Shifts during Covid-19.' Accessed on 15/06/2021 at https://climateoutreach.org/britain-talks-climate/seven-segments-big-picture/shifts-during-covid.

10 Morse, J.W., Gladkikh, T.M., Hackenburg, D.M. and Gould, R.K. (2020) 'COVID-19 and human-nature relationships: Vermonters' activities in nature and associated nonmaterial values during the pandemic.' *PLoS ONE 15*, 12, e0243697.

11 Olsen, H. (2021, 29 January) 'Republicans should embrace an "Operation Warp Speed" for climate change.' *Washington Post*. Accessed on 15/06/2021 at www.washingtonpost.com/opinions/2021/01/29/republicans-operation-warp-speed-climate-change.

12 Thibaut, F. and van Wijngaarden-Cremers, P. (2020) 'Women's mental health in the time of Covid-19 pandemic.' *Frontiers in Global Women's Health 1*, 17. Accessed on 15/06/2021 at www.frontiersin.org/articles/10.3389/fgwh.2020.588372/full.

13 Centers for Disease Control and Prevention (2020, 10 December) 'COVID-19 racial and ethnic health disparities.' Accessed on 15/06/2021 at www.cdc.gov/coronavirus/2019-ncov/community/health-equity/racial-ethnic-disparities/index.html.

14 BBC News (2021, 20 March) 'Covid: Rich states "block" vaccine plans for developing nations.' Accessed on 15/06/2021 at www.bbc.co.uk/news/world-56465395.

15 Barbière, C. (2020, 13 March) 'Coronavirus: Macron announces drastic measures in France.' *Euractiv*. Accessed on 15/06/2021 at www.euractiv.com/section/coronavirus/news/france-to-close-schools-to-curb-coronavirus-spread.

16 Charly Cox said this during a wonderful and inspiring workshop, which we couldn't reference directly; instead, we encourage you to learn more about her work at www.climatechangecoaches.com.

17 Bastida, X. (2020) 'Calling In.' In A.E. Johnson and K.K. Wilkinson (eds) *All We Can Save: Truth, Courage, and Solutions for the Climate Crisis*. New York, NY: One World, p.7.

Chapter 2

1 McGinn, M. (2019, 27 December) '2019's biggest pop-culture trend was climate anxiety.' *Grist*. Accessed on 15/06/2021 at https://grist.org/politics/2019s-biggest-pop-culture-trend-was-climate-anxiety.

2 Panu, P. (2020) 'Anxiety and the ecological crisis: An analysis of eco-anxiety and climate anxiety.' *Sustainability 12*, 9. Accessed on 15/06/2021 at www.researchgate.net/publication/344363346_Anxiety_and_the_Ecological_Crisis_An_Analysis_of_Eco-Anxiety_and_Climate_Anxiety.

3 Amnesty International (2019, 10 December) 'Climate change ranks highest as vital issue of our time – Generation Z survey.' Accessed on 15/06/2021 at www.amnesty.org/en/latest/news/2019/12/climate-change-ranks-highest-as-vital-issue-of-our-time.

4 Washington Post (2020, 3 February) 'The environmental burden of Generation Z.' *The Washington Post Magazine*. Accessed on
15/06/2021 at www.washingtonpost.com/magazine/2020/02/03/
eco-anxiety-is-overwhelming-kids-wheres-line-between-education-alarmism.

5 Triodos Bank (2019, 8 July) 'How is the environmental emergency making us feel?' Accessed on 15/06/2021 at www.triodos.co.uk/articles/2019/
how-is-the-environmental-crisis-making-us-feel.

6 Millennium Kids (n.d.) 'Climate Change.' Accessed on 15/06/2021 at www.
millenniumkids.com.au/the-issues/climate-change.

7 United Nations Development Programme (UNDP) (2021) *Peoples' Climate
Vote*. Accessed on 15/06/2021 at www.undp.org/content/undp/en/home/
librarypage/climate-and-disaster-resilience-/The-Peoples-Climate-Vote-Re-
sults.html.

8 Clayton, S. and Karazsia, B.T. (2020) 'Development and validation of a measure
of climate change anxiety.' *Journal of Environmental Psychology 69*, 101434.

9 American Psychological Association (2020) 'Majority of US adults believe climate change is most important issue today.' Accessed on 15/06/2021 at www.
apa.org/news/press/releases/2020/02/climate-change.

10 Panu, P. (2020) 'Anxiety and the ecological crisis: An analysis of eco-anxiety
and climate anxiety.' *Sustainability 12*, 9. Accessed on 15/06/2021 at www.
researchgate.net/publication/344363346_Anxiety_and_the_Ecological_Cri-
sis_An_Analysis_of_Eco-Anxiety_and_Climate_Anxiety.

11 American Psychiatric Association (2013) *Diagnostic and Statistical Manual of
Mental Disorders* (5th ed.). Washington, DC: APA.

12 Morganstein, J.C. and Ursano, R.J. (2020) 'Ecological disasters and mental
health: Causes, consequences, and interventions.' *Frontiers in Psychiatry 11*.
doi: 10.3389/fpsyt.2020.00001.

13 Cianconi, P., Betrò, S. and Janiri, L. (2020) 'The impact of climate change on
mental health: A systematic descriptive review.' *Frontiers in Psychiatry 11*. doi:
10.3389/fpsyt.2020.00074.

14 Nellemann, C., Verma, R. and Hislop, L. (2011) *Women at the Frontline of Climate Change: Gender Risks and Hopes. A Rapid Response Assessment*. Arendal,
Norway: United Nations Environment Programme, GRID-Arendal.

15 Bartlett, S. (2008) 'Climate change and urban children: Impacts and implications for adaptation in low- and middle-income countries.' *Environment and
Urbanization 20*, 2. doi: 10.1177%2F0956247808096125.

16 Mindlis, I. and Boffetta, P. (2017) 'Mood disorders in first- and second-gen-
eration immigrants: Systematic review and meta-analysis.' *British Journal of
Psychiatry 210*, 3, 182–189.

17 Tschakert, P., Tutu, R. and Alcaro, A. (2013) 'Embodied experiences of environmental and climatic changes in landscapes of everyday life in Ghana.' *Emotion,
Space and Society 7*, 13–25.

18 Gibson, K.E., Barnett, J., Haslam, N. and Kaplan, I. (2020) 'The mental health impacts of climate change: Findings from a Pacific Island atoll nation.' *Journal of Anxiety Disorders 73*, 102237.

19 Chen, S., Bagrodia, R., Pfeffer, C.C., Meli, L. and Bonanno, G.A. (2020) 'Anxiety and resilience in the face of natural disasters associated with climate change: A review and methodological critique.' *Journal of Anxiety Disorders 76*, 102297.

20 Aldrich, D.P. and Meyer, M.A. (2015) 'Social capital and community resilience.' *American Behavioral Scientist 59*, 2. doi: 10.1177%2F0002764214550299.

21 BBC Newsround (2020, 3 March) 'Climate anxiety: Survey for BBC Newsround shows children losing sleep over climate change and the environment.' Accessed on 15/06/2021 at www.bbc.co.uk/newsround/51451737.

22 Wu, J., Snell, G. and Samji, H. (2020) 'Climate anxiety in young people: A call to action.' *The Lancet Planetary Health 4*, 10, E435–E456.

23 Reinhart, R.J. (2018, 11 May) 'Global warming age gap: Younger Americans most worried.' *Gallup*. Accessed on 15/06/2021 at https://news.gallup.com/poll/234314/global-warming-age-gap-younger-americans-worried.aspx.

24 Hickman, C., Marks, E., Pihkala, P., Clayton, S. *et al.* (2021) 'Young People's Voices on Climate Anxiety, Government Betrayal and Moral Injury: A Global Phenomenon.' *Lancet Planetary Health*. Advance online publication. Available at SSRN: https://ssrn.com/abstract=3918955.

25 Twenge, J.M., Cooper, A.B., Joiner, T.E., Duffy, M.E. and Binau, S.G. (2019) 'Age, period, and cohort trends in mood disorder indicators and suicide-related outcomes in a nationally representative dataset, 2005–2017.' *Journal of Abnormal Psychology 128*, 3, 185–199.

26 Washington Post (2020, 3 February) 'The environmental burden of Generation Z.' Accessed on 15/06/2021 at www.washingtonpost.com/magazine/2020/02/03/eco-anxiety-is-overwhelming-kids-wheres-line-between-education-alarmism.

27 Clayton, S. and Karazsia, B.T. (2020) 'Development and validation of a measure of climate change anxiety.' *Journal of Environmental Psychology 69*, 101434.

28 Duggan, J. (n.d.) 'Is this how you feel?' Accessed on 15/06/2021 at www.isthishowyoufeel.com.

29 Jones, M.K., Wootton, B.M., Vaccaro, L.D. and Menzies, R.G. (2012) 'The impact of climate change on obsessive compulsive checking concerns.' *Australian & New Zealand Journal of Psychiatry 46*, 3, 265–270.

30 Clayton, S. (2020) 'Climate anxiety: Psychological responses to climate change.' *Journal of Anxiety Disorders 74*, 102263.

31 Maslin, M. (2021) *How to Save Our Planet: The Facts*. London: Penguin Random House.

32 Nunez, C. (2019, 22 January) 'What is global warming, explained.' *National Geographic*. Accessed on 15/06/2021 at www.nationalgeographic.com/environment/article/global-warming-overview.

33 Jones, S. (n.d.) 'How is climate change affecting the weather today?' MIT Climate Portal. Accessed on 15/06/2021 at https://climate.mit.edu/ask-mit/how-climate-change-affecting-weather-today.

34 Smithsonian National Museum of Natural History (n.d.) 'Extinction Over Time.' Accessed on 15/06/2021 at http://naturalhistory.si.edu/education/teaching-resources/paleontology/extinction-over-time.

35 Smil, V. (2012) *Harvesting the Biosphere: What We Have Taken from Nature.* Cambridge, MA: MIT Press.

36 The Nature Conservancy (2020, 3 November) '2020 Ballot Measures: Seven States, $2.2 Billion for Conservation.' Accessed on 15/06/2021 at www.nature.org/en-us/about-us/who-we-are/how-we-work/policy/state-ballot-initiatives.

37 Dudley, B. (2019, 11 June) 'BP Statistical Review of World Energy.' *British Petroleum.* Accessed on 15/06/2021 at www.bp.com/content/dam/bp/business-sites/en/global/corporate/pdfs/news-and-insights/speeches/bp-stats-review-2019-bob-dudley-speech.pdf.

38 Rapier, R. (2020, 2 August) 'Renewable energy growth continues at a blistering pace.' *Forbes.* Accessed on 15/06/2021 at www.forbes.com/sites/rrapier/2020/08/02/renewable-energy-growth-continues-at-a-blistering-pace.

39 Gibbens, S. (2020, 7 December) '7 wins that gave us hope for the environment in 2020.' Accessed on 15/06/2021 at www.nationalgeographic.com/environment/article/seven-wins-for-the-environment-in-2020.

40 Campbell, P. and Miller, J. (2021, 30 January) 'Electric cars surge in popularity after manufacturers' late dash.' *Financial Times.* Accessed on 15/06/2021 at www.ft.com/content/e9a6aa4f-4a8b-4c80-a89b-13e8fbde2c43.

41 Great Green Wall. (n.d.) 'The Great Green Wall: Growing a world wonder.' Accessed on 15/06/2021 at www.greatgreenwall.org/about-great-green-wall.

42 United Nations Convention to Combat Desertification (2021, 11 January) 'Great Green Wall receives over $14 billion to regreen the Sahel – France, World Bank listed among donors.' Accessed on 15/06/2021 at www.unccd.int/news-events/great-green-wall-receives-over-14-billion-regreen-sahel-france-world-bank-listed-0.

43 Al Jazeera (2020, 29 April) 'Pakistan's virus-idled workers hired to plant trees.' Accessed on 15/06/2021 at www.aljazeera.com/news/2020/4/29/pakistans-virus-idled-workers-hired-to-plant-trees.

44 Brzezinski, K. (2020, December 22) 'Good news! 31 Positive Environmental Stories from 2020.' Accessed on 15/06/2021 at https://onetreeplanted.org/blogs/stories/good-news-2020.

45 Fridays for Future (n.d.) 'Strike statistics.' Accessed on 15/06/2021 at https://fridaysforfuture.org/what-we-do/strike-statistics.

46 Lotzof, K. (n.d.) 'Are we really made of stardust?' Natural History Museum. Accessed on 15/06/2021 at www.nhm.ac.uk/discover/are-we-really-made-of-stardust.html.

47 Maslin, M. (2021) *How to Save Our Planet: The Facts.* London: Penguin Random House.

Chapter 3

1   Adapted from www.psychologytools.com.
2   Adapted from Clayton, S. and Karazsia, B.T. (2020) 'Development and val-idation of a measure of climate change anxiety.' *Journal of Environmental Psychology 69*, 101434. Used with permission.

Chapter 4

1   Nguyen, Q. (2019, 25 March) 'Why fear and anger are rational responses to climate change.' *The Conversation*. Accessed on 15/06/2021 at https://theconversation.com/why-fear-and-anger-are-rational-responses-to-climate-change-114056.
2   Brosch, T. (2021) 'Affect and emotions as drivers of climate change perception and action: A review.' *Current Opinion in Behavioral Sciences 42*, 15–21.
3   Albrecht, G.A. (2019) *Earth Emotions: New Words for a New World*. Ithaca, NY: Cornell University Press.
4   Connell, J.H. (1978) 'Diversity in tropical rain forests and coral reefs.' *Science 199*, 4335, 1302–1310.
5   Singer, E. (2015, 12 February) 'Game theory calls cooperation into question.' *QuantaMagazine*. Accessed on 15/06/2021 at www.quantamagazine.org/game-theory-explains-how-cooperation-evolved-20150212.
6   Cunsolo, A. and Ellis, N.R. (2018) 'Ecological grief as a mental health response to climate change-related loss.' *Nature Climate Change 8*, 4, 275–281.
7   Kübler-Ross, E. (1969) *On Death and Dying*. New York, NY: Scribner.
8   Randall, R. (2009) 'Loss and climate change: The cost of parallel narratives.' *Ecopsychology 1*, 3, 118–129, p.121.
9   Ibid., p.124.
10 Ibid., p.119.
11 Stoknes, P.E. (2015) *What We Think About When We Try Not to Think About Global Warming: Toward a New Psychology of Climate Action*. White River Junction, VT: Chelsea Green Publishing, p.176.
12 Grist (2020, 8 April) 'This professor wants you to give up your climate guilt.' Accessed on 15/06/2021 at https://grist.org/fix/this-professor-wants-you-to-give-up-your-climate-guilt.

Chapter 5

1   Butcher, J. (2009) *The Dresden Files: Turn Coat*. Waterville, ME: Thorndike Press.

2 Lertzman, R. (2015) *Environmental Melancholia: Psychoanalytic Dimensions of Engagement*. London: Routledge.

3 This quote is widely attributed to Carl Jung, but no direct reference applies. Perhaps the best single example of his work on the unconscious is the essay 'The Relations Between the Ego and the Unconscious' in: Jung, C.G. (1967) *Two Essays on Analytical Psychology, Collected Works of C.G. Jung, Volume 7*. Princeton, NJ: Princeton University Press.

4 Bendell, J. (2018) 'Deep adaptation: A map for navigating climate tragedy.' Accessed on 30/06/2021 at https://jembendell.com.

5 Humphrys, J. (2019, 17 April) 'Extinction Rebellion: Noble and Necessary or a Pointless Nuisance?' Accessed on 15/06/2021 at https://yougov.co.uk/topics/science/articles-reports/2019/04/17/john-humphrys-extinction-rebellion-noble-and-neces.

6 Tsjeng, Z. (2019, 27 February) 'The Climate Change Paper So Depressing It's Sending People to Therapy.' Accessed on 15/06/2021 at www.vice.com/en/article/vbwpdb/the-climate-change-paper-so-depressing-its-sending-people-to-therapy.

7 Stoknes, P.E. (2015) *What We Think About When We Try Not to Think About Global Warming: Toward a New Psychology of Climate Action*. White River Junction, VT: Chelsea Green Publishing, p.82.

8 Festinger, L., Riecken, H.W. and Schachter, S. (1956) *When Prophecy Fails*. Minneapolis, MN: University of Minnesota Press, p.3.

9 DutchNews.nl (2021, 15 February) 'Avocados more popular than ever but at an environmental cost.' Accessed on 15/06/2021 at www.dutchnews.nl/news/2021/02/avocados-more-popular-than-ever-but-at-an-environmental-cost.

10 Nature (2010, 17 November) 'Closing the Climategate.' *Nature 468*, 345. Accessed on 15/06/2021 at www.nature.com/articles/468345a.

11 Lewandowsky, S. and Oberauer, K. (2016) 'Motivated rejection of science.' *Current Directions in Psychological Science 25*, 4, 217.

12 Stoknes, P.E. (2015) *What We Think About When We Try Not to Think About Global Warming: Toward a New Psychology of Climate Action*. White River Junction, VT: Chelsea Green Publishing, p.74.

13 Kahneman, D. (2011) *Thinking, Fast and Slow*. New York, NY: Farrar, Straus & Giroux.

14 The Center for Research on Environmental Decisions produced, as early as 2009, an excellent and succinct report, which still serves as a useful primer for how biases in thinking affect how we receive climate messages: Center for Research on Environmental Decisions (2009) *The Psychology of Climate Change Communication: A Guide for Scientists, Journalists, Educators, Political Aides, and the Interested Public*. New York, NY: CRED.

15 Bell, J. (2020) *Radical Attention*. London: Peninsula Press. Cited in Cosslett, R.L. (2020) 'If you find yourself doomscrolling your way through social media, help is at hand.' *The Guardian*, 19 October. Accessed on 30/06/2021 at www.theguardian.com/commentisfree/2020/oct/19/doomscrolling-social-media-apocalyptic-updates-cure.

16 Fitzgerald, S. (2020, 30 July) 'The Internet wants to keep you "doom-scrolling". Here's how to break free.' *Washington Post*. Accessed on 15/06/2021 at www.washingtonpost.com/lifestyle/wellness/coronavirus-doom-scrolling-stop/2020/07/29/2c87e9b2-d034-11ea-8d32-1ebf4e9d8e0d_story.html.

17 Barnidge, M. (2018) 'Social affect and political disagreement on social media.' *Social Media + Society 4*, 3. Accessed on 15/06/2021 at https://journals.sagepub.com/doi/pdf/10.1177/2056305118797721.

Chapter 6

1 Gore, A. (2006) *An Inconvenient Truth: The Planetary Emergency of Global Warming and What We Can Do About It.* New York, NY: Rodale.

2 Details of the Yale Programme on Climate Communication, including the online courses, can be found at: https://climatecommunication.yale.edu.

3 Wang, S., Corner, A. and Nicholls, J. (2020) *Britain Talks Climate: A Toolkit for Engaging the British Public on Climate Change*. Oxford: Climate Outreach.

4 Hassol, S.J. (2015, June) 'ClimateTalk: Science and Solutions (Video).' *TEDxUMontana*. Accessed on 15/06/2021 at https://steamregister.com/climatetalk-science-and-solutions-susan-joy-hassol-tedxumontana.

5 Lacroix, K., Goldberg, M.H., Gustafson, A., Rosenthal, S.A. and Leiserowitz, A. (2020) 'Should it be called "natural gas" or "methane"?' *Yale Program on Climate Change Communication*. Accessed on 30/06/2021 at https://climatecommunication.yale.edu/publications/should-it-be-called-natural-gas-or-methane.

6 Goldberg, M., Gustafson, A., Rosenthal, S., Kotcher, J., Maibach, E. and Leiserowitz, A. (2020) 'For the first time, the Alarmed are now the largest of Global Warming's Six Americas.' *Yale Program on Climate Change Communication*. Accessed on 30/06/2021 at https://climatecommunication.yale.edu/publications/for-the-first-time-the-alarmed-are-now-the-largest-of-global-warmings-six-americas.

7 United Nations Development Programme (2021) *The Peoples' Climate Vote: Results*. Oxford: UNDP. Accessed on 15/06/2021 at www.undp.org/publications/peoples-climate-vote.

8 The Peoples' Climate Vote is so rich in information it was challenging to know what to include in the main body, but we encourage looking at the entire report.

9 Climate Outreach (n.d.) 'Seven British segments: The big picture.' Accessed on 15/06/2021 at https://climateoutreach.org/britain-talks-climate/seven-segments-big-picture.

10 Climate Outreach (n.d.) 'Common ground and differences on climate change.' Accessed on 15/06/2021 at https://climateoutreach.org/britain-talks-climate/seven-segments-big-picture/common-ground-differences.

11 Interview with Dr Adam Corner, research director of Climate Visuals. Accessed on 15/06/2021 at https://medium.com/wedonthavetime/climate-change-had-an-image-problem-and-we-set-out-to-change-that-8635a5098e84.

12 Smith, T. (2021, 24 February) 'Climate Visuals – proving that imagery needs to embody people-centred narratives and positive solutions.' Accessed on 15/06/2021 at https://climateoutreach.org/climate-visuals-proving-imagery-needs-people-centred-narratives-solutions.

13 We would really encourage people to take a look at the Climate Visuals project, which is developing an evidence-based photography resource to help shift the 'image problem' in climate change, available at: https://climatevisuals.org.

14 Climate Psychology Alliance (2020, 20 September) 'Handbook of Climate Psychology.' Accessed on 15/06/2021 at www.climatepsychologyalliance.org/handbook.

15 Bonneau, A.-M. (2021) *The Zero-Waste Chef: Plant-Forward Recipes and Tips for a Sustainable Kitchen and Planet*. New York, NY: Avery Publishing.

16 Webster, R. and Marshall, G. (2019) *The #TalkingClimate Handbook: How to Have Conversations about Climate Change in Your Daily Life*. Oxford: Climate Outreach, p.6.

17 Stoknes, P.E. (2015) *What We Think About When We Try Not to Think About Global Warming: Toward a New Psychology of Climate Action*. White River Junction, VT: Chelsea Green Publishing.

18 American Psychological Association (n.d.) 'Narrative psychology – APA Dictionary of Psychology.' Accessed on 15/06/2021 at https://dictionary.apa.org/narrative-psychology.

19 Norton-Smith, K., Lynn, K., Chief, K., Cozzetto, K. *et al.* (2016) 'Climate change and indigenous peoples: A synthesis of current impacts and experiences.' Gen. Tech. Rep. PNW-GTR-944. Portland, OR: US Department of Agriculture, Forest Service, Pacific Northwest Research Station.

20 Ibid.

Chapter 7

1 Lertzman, R. (2020, March) 'How to turn climate anxiety into action.' *TED Talk (Video)*. Accessed on 15/06/2021 at www.ted.com/talks/renee_lertzman_how_to_turn_climate_anxiety_into_action/transcript?language=en.

2 Gilbert, P. (2010) 'Training Our Minds in, with and for Compassion: An Introduction to Concepts and Compassion-Focused Exercises.' Accessed on 15/06/2021 at www.getselfhelp.co.uk/docs/GILBERT-COMPASSION-HANDOUT.pdf, p.8.

3   We strongly encourage you to watch the, admittedly stylistically dated, documentary by Bill Moyers, *Healing From Within*, showcasing the incredible mindfulness-based stress reduction (MBSR) programme developed by Jon Kabat-Zinn, at the University of Massachusetts medical centre. At time of writing, this was available at: https://vimeo.com/39767361.

4   Luders, E., Cherbuin, N. and Kurth, F. (2015) 'Forever young(er): Potential age-defying effects of long-term meditation on gray matter atrophy.' *Frontiers in Psychology 5*, 1551.

5   Hölzel, B.K., Carmody, J., Vangel, M., Congleton, C. *et al.* (2011) 'Mindfulness practice leads to increases in regional brain gray matter density.' *Psychiatry Research: Neuroimaging 191*, 1, 36.

6   For a practical guide to mindfulness, we recommend the following book: Williams, M. and Penman, D. (2011) *Mindfulness: A Practical Guide to Finding Peace in a Frantic World*. London: Piatkus.

7   Manson, M. (2016) *The Subtle Art of Not Giving a F*ck: A Counterintuitive Approach to Living a Good Life*. New York, NY: HarperCollins, p.34.

8   Bandura, A. (1977) 'Self-efficacy: Toward a unifying theory of behavioral change.' *Psychological Review 84*, 2, 191–215.

9   Bostrom, A., Hayes, A.L. and Crosman, K.M. (2019) 'Efficacy, action, and support for reducing climate change risks.' *Risk Analysis 39*, 4, 805–828.

10 Bandura, A. (2007) 'Impeding ecological sustainability through selective moral disengagement.' *International Journal of Innovation and Sustainable Development 2*, 1, 8.

11 Burnham, M. and Ma, Z. (2017) 'Climate change adaptation: Factors influencing Chinese smallholder farmers' perceived self-efficacy and adaptation intent.' *Regional Environmental Change 17*, 1, 171–186.

12 Wang, S., Leviston, Z., Hurlstone, M., Lawrence, C. and Walker, I. (2018) 'Emotions predict policy support: Why it matters how people feel about climate change.' *Global Environmental Change 50*, 25–40.

13 © Copyright 2018, VIA Institute on Character, www.viacharacter.org. Used with permission.

14 Nature-deficit disorder is described by Richard Louv in the following book: Louv, R. (2008) *Last Child in the Woods: Saving our Children from Nature-Deficit Disorder*. Chapel Hill, NC: Algonquin Books.

15 See the following for a detailed global mapping review of nature-based interventions: Shanahan, D.F., Astell-Burt, T., Barber, E.A., Brymer, E. *et al.* (2019) Nature-based interventions for improving health and wellbeing: The purpose, the people and the outcomes. *Sports 7*, 6, 141.

16 Details can be found at the OCR GCSE Natural History Hub website at: https://teach.ocr.org.uk/naturalhistory.

17 Durrell, G. (2006) *The Corfu Trilogy: My Family and Other Animals; Birds, Beasts and Relatives; and The Garden of the Gods*. London: Penguin.

18 This book is simply incredible: McAnulty, D. (2020) *Diary of a Young Naturalist*. London: Penguin.

Chapter 8

1   Climate Psychology Alliance (2020, 20 September) 'Handbook of Climate Psychology.' Accessed on 15/06/2021 at www.climatepsychologyalliance.org/handbook.

2   Maslin, M. (2021) *How to Save Our Planet: The Facts.* London: Penguin Random House. We can't recommend Maslin's book highly enough. It breaks down, in simple and accessible ways, the science of climate change and what we can do as individuals at home, at work, in the community and in our wider systems.

3   The Carbon Literacy Project offers certification to individuals and organizations regarding how to become more aware of the carbon costs of everyday activity, and how best to take actions to reduce its impacts. It can be accessed at: https://carbonliteracy.com.

4   Hastings, C. (2020) *The Carbon Buddy Manual.* Truro: The Carbon Buddy Project.

5   Andrus, A. (2018) *101 Small Ways to Change the World.* Victoria, Australia: Lonely Planet Global.

6   Find out more about this pioneering project at https://projectinsideout.net.

7   Gale, J. (2020) *The Sustainable (ish) Living Guide: Everything You Need to Know to Make Small Changes That Make a Big Difference.* Newnan, GA: Green Tree.

8   Gale, J. (2021) *The Sustainable(ish) Guide to Green Parenting: Guilt-free Eco-ideas for Raising Your Kids.* Newnan, GA: Green Tree.

9   Gale, J. (n.d.) 'About Sustainable(ish).' Accessed on 15/06/2021 at www.asustainablelife.co.uk/about-2.

10 Thaler, R.H. and Sunstein, C.R. (2009) *Nudge: Improving Decisions About Health, Wealth, and Happiness.* London: Penguin.

11 Ibid., p.3.

12 Afif, Z. (2017, 25 October) '"Nudge units" – where they came from and what they can do.' World Bank Blogs. Accessed on 15/06/2021 at https://blogs.worldbank.org/developmenttalk/nudge-units-where-they-came-and-what-they-can-do.

13 Whitehead, M., Jones, R., Howell, R., Lilley, R. and Pykett, J. (2014) *Nudging all over the World: Assessing the Global Impact of the Behavioural Sciences on Public Policy.* ESRC Report, Swindon: Economic and Social Research Council.

14 Schubert presented an excellent round-up of Green Nudges, as well as examined whether they were effective and ethically justifiable: Schubert, C. (2017) 'Green nudges: Do they work? Are they ethical?' *Ecological Economics 132,* 329–342.

15 Rozin, P., Scott, S.E., Dingley, M., Urbanek, J.K., Jiang, H. and Kaltenbach, M. (2011) 'Nudge to nobesity I: Minor changes in accessibility decrease food intake.' *Judgement and Decision Making 6,* 4, 323–332.

16 Kallbekken, S. and Sælen, H. (2013) '"Nudging" hotel guests to reduce food waste as a win–win environmental measure.' *Economics Letters 119,* 3, 325–327.

17 Holden, S.S., Zlatevska, N. and Dubelaar, C. (2016) 'Whether smaller plates reduce consumption depends on who's serving and who's looking: A meta-analysis.' *Journal of the Association for Consumer Research 1*, 1, 134.

18 Robinson, E., Nolan, S., Tudur-Smith, C., Boyland, E.J. *et al.* (2014) 'Will smaller plates lead to smaller waists? A systematic review and meta-analysis of the effect that experimental manipulation of dishware size has on energy consumption.' *Obesity Reviews 15*, 10, 812–821.

19 Livingstone, M.B.E. and Pourshahidi, L.K. (2014) 'Portion size and obesity.' *Advances in Nutrition 5*, 6, 829–834.

20 Letwin, O., Barker, G. and Stunell, A. (2011) *Behaviour Change and Energy Use.* London: Cabinet Office: Behavioural Insights Team.

21 Pichert, D. and Katsikopoulos, K.V. (2008) Green defaults: Information presentation and pro-environmental behaviour. *Journal of Environmental Psychology 28*, 1, 63.

22 Egebark, J. and Ekström, M. (2016) 'Can indifference make the world greener?' *Journal of Environmental Economics and Management 76*, C, 1–13.

23 Wray, B. (2020, 5 August) 'Why activism isn't *really* the cure for eco-anxiety and eco-grief.' Accessed on 15/06/2021 at https://gendread.substack.com/p/why-activism-isnt-really-the-cure.

24 Gale, J. (n.d.) 'Sustainable(ish).' Accessed on 15/06/2021 at www.asustainablelife.co.uk/page/2/?pages-list.

25 Coaches love scheduling. A number of prominent coaches specifically highlight the benefits of scheduling as a form of self-care, facilitating goal achievement as well as the importance of valuing your time and input. For more amazing in-depth wisdom on scheduling, see: Kim John Payne (founder of the Simplicity Parenting programme), https://www.simplicityparenting.com; Co-Active Training Institute, https://coactive.com/blog/setting-yourself-up-for-agile-leadership-in-todays-chaotic-world; Brooke Castillo (author of *Self Coaching 101* and founder of The Life Coach School and *The Life Coach School Podcast*), https://thelifecoachschool.com; Clare Evans (*Time Management for Dummies*), https://clareevans.co.uk/create-meaningful-goals.

26 Brown, B. [@BreneBrown] (2020, 24 April) When we hit that wall, sometimes courage looks like scaling it or breaking through it. AND, sometimes courage is building a fort against the wall and taking a nap. Hard days are real because this is hard. Stay awkward, kind and brave enough to rest and feel. [Tweet]. Twitter. Accessed on 15/06/2021 at https://twitter.com/BreneBrown/status/1253720039506817025.

27 4ocean (n.d.) 'Total recovered since 2017: 15,7111,665 lbs.' Accessed on 20/06/2021 at www.4ocean.com/pages/our-impact.

28 Howard, B.C. (2017, July) 'Al Gore: The green revolution is "unstoppable".' *National Geographic*. Accessed on 15/06/2021 at www.nationalgeographic.com/magazine/article/3-questions-al-gore-climate-change.

Further Reading

Attenborough, D. (1980) *Life on Earth*. London: BBC Books and William Collins Sons & Co.

Gates, B. (2021) *How to Avoid a Climate Disaster: The Solutions We Have and the Breakthroughs We Need*. New York, NY: Alfred A. Knopf.

Grose, A. (2020) *A Guide to Eco-Anxiety: How to Protect the Planet and Your Mental Health*. London: Watkins Media Limited.

Johnson, J.P. (2007) *The Armchair Naturalist: How to Be Good at Nature Without Really Trying*. London: Icon Books.

Tree, I. (2018) *Wilding: The Return of Nature to a British Farm*. London: Picador.

Index